Simple Birdhouses and Feeders of Wood

Ronald D. Tarjany

tp
Tarjany Publications
PO Box 8846 Calabasas CA 91302

Copyright © 2000 by Tarjany Publications

This book may not be reproduced, either in part or in its entirety, in any form, by any means, without permission from the publisher, with the exception of brief exerpts for the purpose of radio, television, or published reviews. Although all possible measures have been taken to ensure the accuracy of the material presented, neither the author nor Tarjany Publications is liable in case of misinterpretation of directions, misapplication or typographical error. All rights, including the right of translation, are reserved.

Printed in United States of America

10 9 8 7 6 5 4 3 2 1
First Edition

ISBN 0-9674668-3-0

This Book was Produced by
TARJANY PUBLICATIONS
PO Box 8846, Calabasas, CA 91302

Editorial: Cathy Starbird
Copy Editor: Inge Kriegler
Design: Les Ventura
Illustrations: David R. Shea
Photography: T. R. Douglass
Cover photographs: Frank G. Farkas

Library of Congress Cataloging-in-Publication Data

Tarjany, Ronald D.
    Simple Birdhouses and Feeders of Wood

    Includes index.
    ISBN 0-9674668-3-0
    1. Birdhouses---Design and construction.
    2. Bird Feeders---Design and construction.
    I. Title. II. Title: Simple Birdhouses and Feeders of Wood
    Library of Congress Card Number: 00-190449

Contents

Author's Notes 4
PROJECTS
    Typical Birdhouse 5
    Wren Birdhouse 9
    Carolina Wren Tower Birdhouse. 12
    Bluebird Birdhouse. 17
    Western Martin Triplex Birdhouse 21
    Chickadee Birdhouse. 27
    Small Hanging Bird Feeder 31
    Hopper Wall Mounted Bird Feeder 34
    Covered Hopper Bird Feeder. 39
    Open Hanging Bird Feeder 43
    Multi Station Bird Feeder 46
    Center Column Bird Feeder 51
    Bird Water Platform 55
Mounting and Poles. 58
Predator Guards 60
Birdhouse Size Specifications. 62
Index 64

Author's Notes

Of the many hobbies that I have, I probably enjoy woodworking the most. And, of the many facets of woodworking, I must say I like building birdhouses and feeders the best, after building wooden toys. Building a simple birdhouse is easy to do, since fancy cuts and precision accuracy are not that important. Butt joints with water proof glue and a couple of box nails are all that are needed for most simple birdhouses. Once the house is hung from a limb of a tree or put on top of a pole in the backyard, no one will be checking out the workmanship; they will be watching the birds. The birdhouse or the feeder will only be background items to view. For these reasons birdhouse building is ideal for those new persons to start building things from wood. They can hone their skills with each new project.

The twelve birdhouses and feeders and the bird water platform presented in this book were constructed mostly from 1/2" construction grade plywood that was left over from a room addition that I built on to my house. The redwood and western red cedar were cut from fence lumber that was purchased at the local home center. Some of the structures were built many years ago when the thickness was 3/4" and others more recently, with fence material that is only 5/8" thick. Either thickness can be used; just adjust the necessary dimensions to take this into account.

The word "simple" used in the title of this book may cause you concern, when you look at some of the structures in this book, like the multi position bird feeder. You may say this is not a simple project to build! I think it is a simple feeder; it just has more parts than many of the other projects! There are no fancy joints to cut; they are all simple butt joints!

Typical Birdhouse

The typical birdhouse will be well suited for several different types of birds. The roof is hinged at the back for easy cleaning. The birdhouse is made of 1/2" construction grade plywood with the rough side to the inside except for the roof which has the rough side to the outside for a contrasting texture. It is best to have the rough side inside so the birds can cling to the inside wall easier.

Rip a 25" long piece of 1/2" thick plywood 4" wide for the floor (**A**), the front (**B**) and back (**C**). Crosscut a piece 4" long for the floor. Lay out the centers of the four drain holes 3/8" from the edges and drill with a 1/4" diameter bit. Cut the front 9 1/2" long and use the other piece for the back. Put these two pieces aside for now; we will cut the angles on the top edges to match the angle on the sides later.

Rip a 21" long piece of 1/2" thick plywood 5" wide for the two sides (**D**). Lay out the angle at the top of the sides 8 3/16" from the bottom edge for the front and 10" for the back. Cut the two pieces on a band saw and sand the edges smooth.

Set your table saw blade to the same angle as the roof slope on the side and cut the front with the short 8 3/16" dimension on the outside face and the bevel going up to the inside face. Cut the back wall with the 10" dimension on the outside face and the bevel going down to the inside face. Lay out the center for the entrance hole and the perch hole in the front. Drill the entrance hole with a 1 1/4" diameter hole saw, sand the inside surface with a drum sander and break the inside and outside edges with a piece of sandpaper. Drill the 1/4" diameter perch hole. Lay out the centers of the two vent holes in the back. Drill the 3/8" diameter vent holes.

Glue and nail, using the four penny box nails (**I**), the front to the edge of the floor. Then glue and nail the sides to the edge of the floor and the edge of the front. Next, glue and nail the back to the edge of the floor and the back of the sides to the edges of the back.

Lay out the roof (**E**) 5" wide by 6 1/2" long on a piece of 1/2" thick plywood. Cut out the roof and sand the edges.

Lay out and cut the support (**F**) 2 1/2"

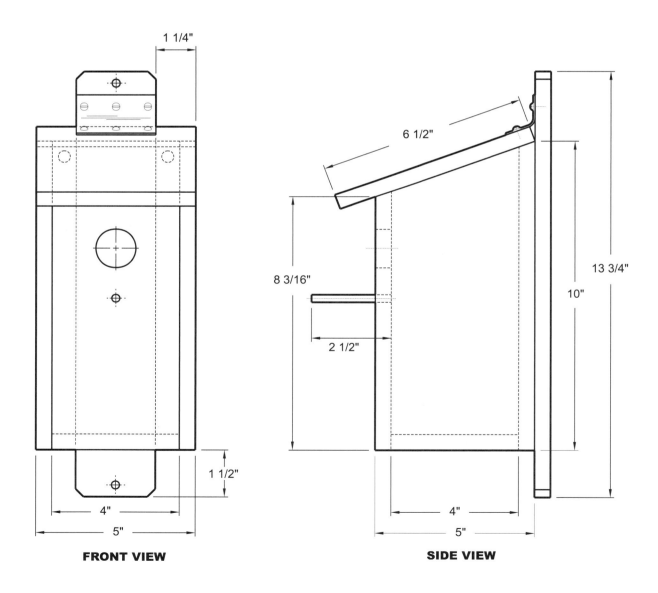

FRONT VIEW　　　　　　　**SIDE VIEW**

Typical Birdhouse

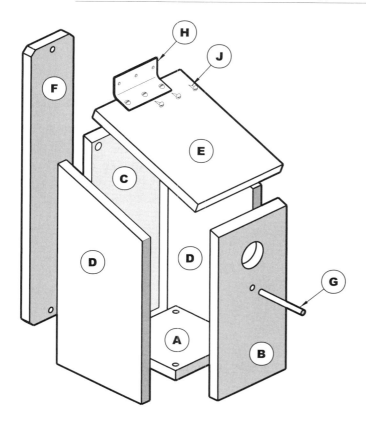

TYPICAL BIRDHOUSE ASSEMBLY

	LIST OF MATERIALS		
	DESCRIPTION	MATERIAL	FINISHED SIZE
A	FLOOR	PLYWOOD	1/2 x 4 x 4
B	FRONT	PLYWOOD	1/2 x 4 x 8 3/8
C	BACK	PLYWOOD	1/2 x 4 x 10
D	SIDE (2)	PLYWOOD	1/2 x 5 x 10
E	ROOF	PLYWOOD	1/2 x 5 x 6 1/2
F	SUPPORT	PLYWOOD	1/2 x 2 1/2 x 13 3/4
G	PERCH	DOWEL	1/4 DIA x 2 1/2
H	HINGE	LEATHER	1/16 x 1 3/4 x 2 1/2
I	BOX NAIL (40)	STEEL	4d GALVANIZED
J	WOOD SCREW (6)	BRASS	1/2 x No 6 RD HEAD

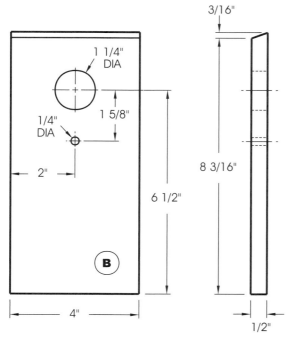

FRONT DETAIL

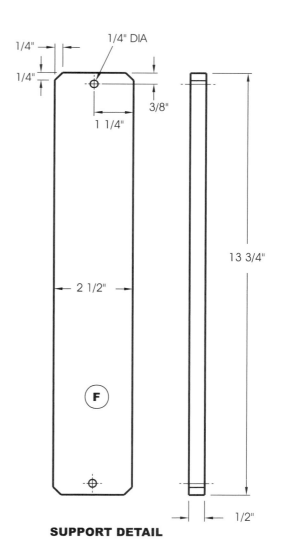

SUPPORT DETAIL

wide by 13 3/4" long on a piece of 1/2" thick plywood. Lay out the four 1/4" by 1/4" chamfers. Cut the chamfers staying away from the line and then sand to the line. Mark the location of the centers for the two mounting holes and drill with a 1/4" drill bit. Glue and nail the support to the back of the birdhouse 1 1/4" from the side and leave 1 1/2" extending past the bottom of the floor.

Use a non-toxic finish for the birdhouse. We used exterior latex enamel to paint our birdhouse. Paint the birdhouse your favorite color and the roof a contrasting color. We painted the exterior walls and the underside exposed part of the roof a light brown and the roof a light green. Do not paint the interior of the house nor the inside of the entrance hole. The perch (**G**) was also left unpainted and will weather over time.

Allow to dry overnight and then glue the perch into the hole in the front wall. Cut the hinge (**H**) 1 3/4" wide by 2 1/2" long from a 1/16" thick piece of leather. Score across the 2 1/2" dimension in the center of the 1 3/4" dimension on one side of the leather with an awl so it will bend easier. Punch holes in the hinge and screw it to the roof and the support, using the 1/2" No.6 round head wood screws (**J**).

Nail the birdhouse to a tree trunk using the two mounting holes in the support. Your first tenants will be checking it out real soon.

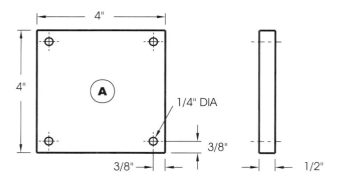

FLOOR DETAIL

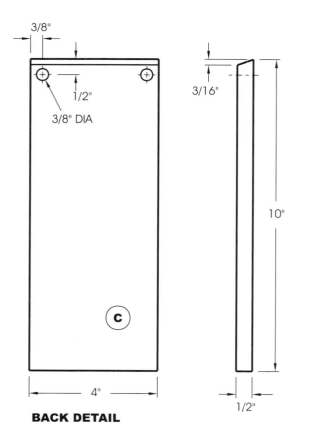

BACK DETAIL

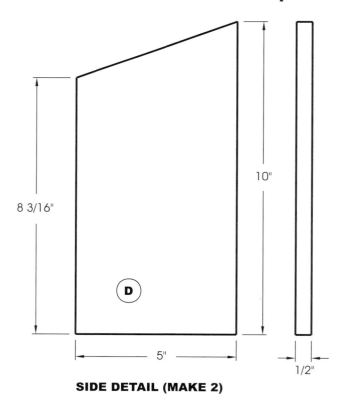

SIDE DETAIL (MAKE 2)

Wren Birdhouse

The diamond shaped birdhouse is a common design that is well suited for the wren. The birdhouse is small and easy to build. The birdhouse is made of 1/2" thick construction grade plywood.

Cut all the pieces to overall size as shown in the list of materials. Lay out the centers for the three holes in the front (**A**) using the dimensions from the front detail drawing. Drill the 1" diameter hole for the entrance, the 1/4" diameter hole for the perch and the 7/32" diameter hole for the screw that will attach the roof to the house.

Lay out the centers for the three holes in the back (**B**) using the dimensions from the back detail drawing. Drill the two 1/4" diameter holes for ventilation and the 7/32" diameter hole for the other screw that will attach the roof to the house.

Cut the 45 degree bevel on the lower edge of the two sides (**C**). Lay out the centers and drill the 1/4" diameter drain holes.

The 3/4" by 3/4" roof beam (**E**) fits up in the roof peak between the front and back walls. Measure down 3/4" from the top peak of the beam. Draw a line 45 degrees to the edge and sand the bottom edge to the line. The eyebolt end does not protrude below the bottom of the roof beam. We do not want our little bird hitting its head on a sharp object. So to get a nut and

Simple Birdhouses and Feeders of Wood

LIST OF MATERIALS

	DESCRIPTION	MATERIAL	FINISHED SIZE
A	FRONT	PLYWOOD	1/2 x 4 1/2 x 4 1/2
B	BACK	PLYWOOD	1/2 x 4 1/2 x 4 1/2
C	SIDE (2)	PLYWOOD	1/2 x 4 1/2 x 4 1/2
D	ROOF (2)	PLYWOOD	1/2 x 6 x 7 1/2
E	ROOF BEAM	CEDAR	3/4 x 3/4 x 4 1/2
F	PERCH	MAPLE	1/4 DIA x 2 1/2
G	ROOF COVER	TAPE	1 1/2 X 7 1/2
H	EYEBOLT	STEEL	3/16 x 2 GALV.
I	WASHER	STEEL	No 10 GALV.
J	NUT	STEEL	10-32 GALV.
K	WOOD SCR (2)	BRASS	1 1/4 x No 8 RD HD
L	BOX NAIL (6)	STEEL	4d GALVANIZED

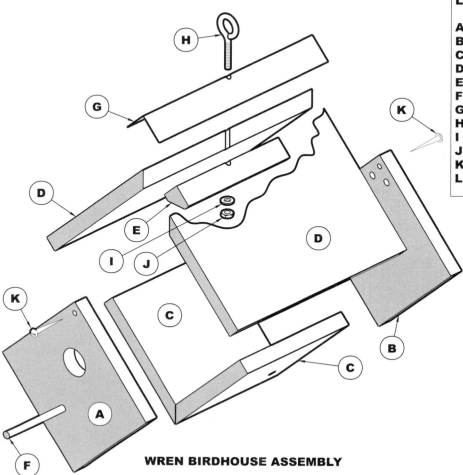

WREN BIRDHOUSE ASSEMBLY

washer on the end of the eyebolt, we must add a counterbore large enough to get at the nut. Mark the center for the counterbore on the underside of the roof beam. Clamp the roof beam so the bottom is facing up and is parallel with the drill press table. Drill the 1/2" diameter hole 1/2" deep using a Forstner bit and then drill a 3/16" diameter hole thru it.

Test fit the sides to the front

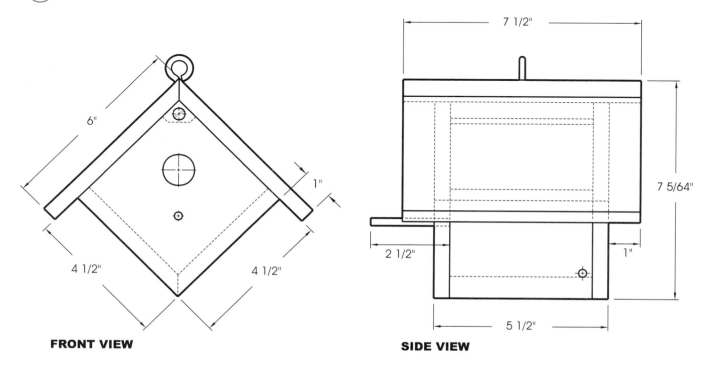

FRONT VIEW　　　　　　　　**SIDE VIEW**

Wren Birdhouse

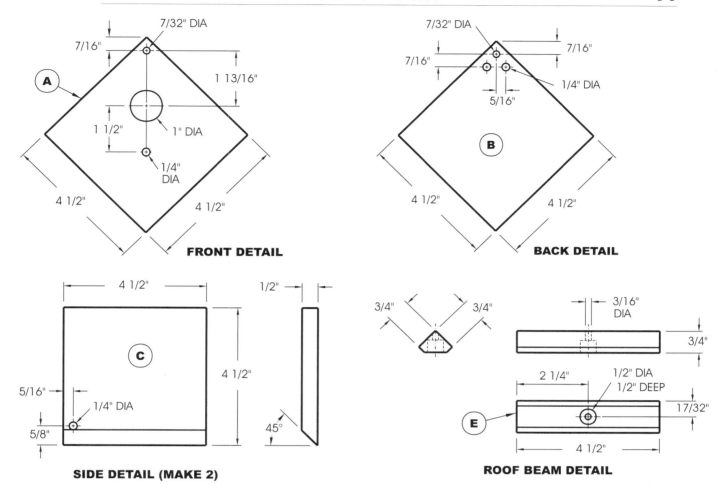

and back walls, and make any adjustments necessary. When everything fits properly glue and nail, using the four penny box nails (**L**), the front and back walls to the edges of the side walls. Use waterproof glue for all joints. Test the fit of the two roof pieces to the house. Clamp one section of the roof to the house and apply glue to the ridge bevel of the other roof section. Join it to the other roof bevel and clamp it to the house. Only the ridge bevel gets glued since the roof needs to be removable for house cleaning. Allow to dry and then remove the clamps. Glue the ridge beam into the underside of the roof with equal clearance to the front and back. When the glue has dried drill a 3/16" diameter hole thru the roof peak from the hole in the roof beam. Turn the roof over and add blocks so the bottom of the roof beam is parallel with your drill press table. Drill the 3/16" diameter hole.

The roof cover (**G**) is a 7/12" long strip of cloth tape 1 1/2' wide. Put half of the tape on one side of the roof at the peak and fold it down the other side. This will help to keep the rain from entering the roof where the two halves join. Pierce the tape with an awl at the eyebolt hole.

Use a non-toxic finish for your birdhouse. Do not paint the interior of the house nor the inside of the entrance hole. We painted our house, roof edges and eves a light brown color and the roof a light green using an exterior latex enamel. Allow to dry overnight and then add the eyebolt (**H**), washer (**I**) and nut (**J**). Put the roof on the birdhouse and add the two round head wood screws (**K**). Glue the perch (**F**) into the hole in the front wall.

Hang your wren birdhouse in a tree and it will not be long before you have your first tenants with some young ones to follow.

Carolina Wren Tower Birdhouse

The Carolina wren tower birdhouse will house three families of birds and can be hung from a branch of a tree. The house is square with a porch for each opening. The entrance holes are on three different sides to give the birds some privacy. The back is removable for easy cleaning of all three homes. The walls are made of 1/2" construction grade plywood with the rough side in and the other pieces are made of 3/4" western red cedar and redwood.

Lay out three floors (**A**) 4" by 4" in line on a piece of 3/4" redwood. Rip the piece of redwood 4" wide and then crosscut the piece to 4" long sections. Use a stop block attached to the tabletop to insure the three pieces are the same size.

Rip two 23" long pieces of 1/2" thick plywood 5" wide for the front (**B**) and back (**C**). Then rip two 23" long pieces of 1/2" thick plywood 4" wide for the right side (**D**) and left side (**E**). Cut the four pieces to a length of 21 3/4". Lay out the center for the entrance and ventilation holes in the front and side pieces. Drill the three entrance holes 1 1/8" diameter and the twenty-four ventilation holes 1/4" diameter. Lay out the centers for the drain holes in the back wall. Drill the 1/4" diameter holes. Lay out the location for the floors on the inside and the location for the trim pieces on the outside of all the walls.

Glue and nail, using the 4 penny box nails (**O**), the edge of the floors to the left side. Then glue and nail the right side to the edge of the floors. Next, glue and nail the front to the edges of the floors and the edges of the sides.

Lay out the top (**F**) and the bottom (**G**) on a piece of 3/4" redwood. Cut the pieces on a band saw and sand the edges smooth. Glue and nail the bottom to the floor. Lay out the center of the hole for the eyebolt in the top. Drill the 7/32" diameter hole.

Lay out the ceiling (**H**) 4" wide by 4" long on a piece of 3/4" thick redwood. Cut out the piece and sand the edges. Mark the center of the clearance hole for the nut. Drill the 3/4" diameter hole. Glue and nail the ceiling edges to the sides and front. Glue and nail the top to the ceiling.

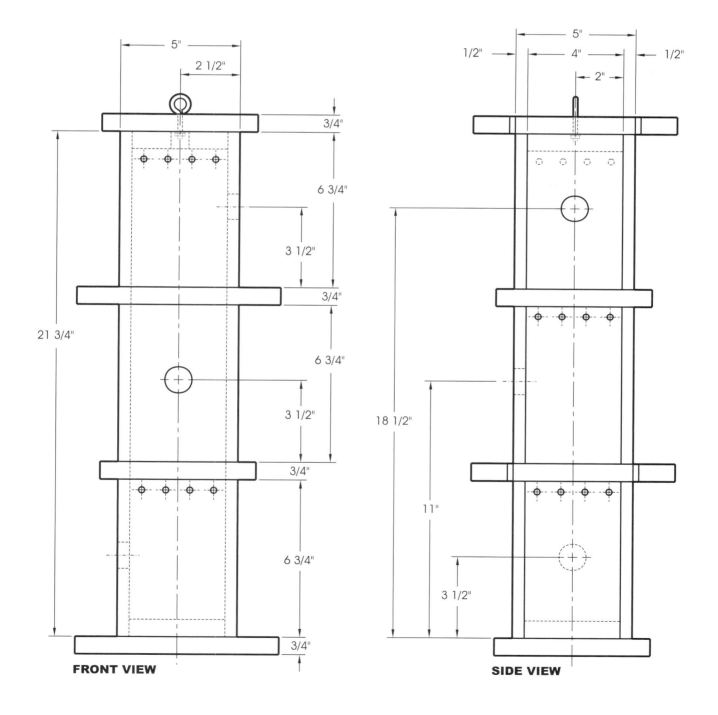

FRONT VIEW　　　　**SIDE VIEW**

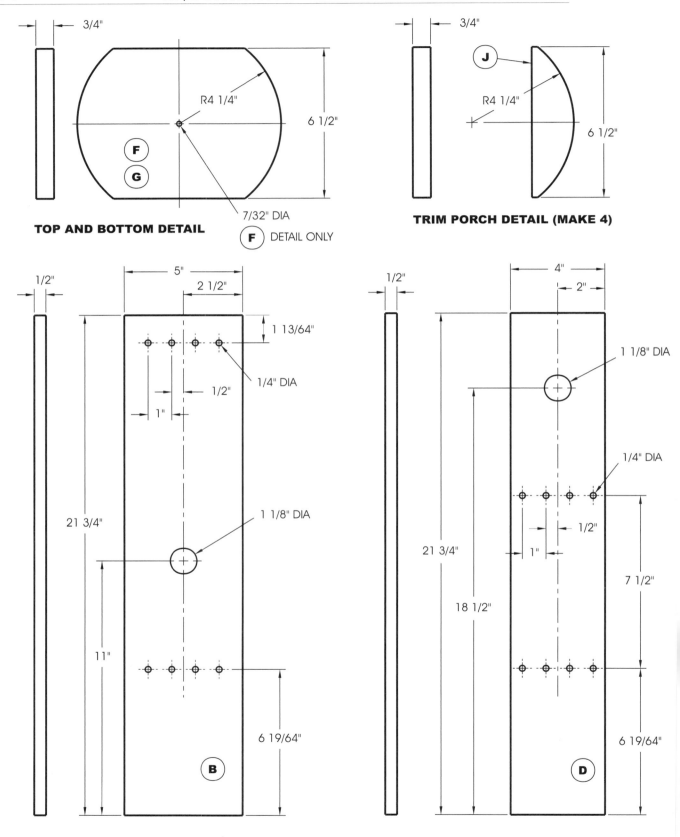

Rip a 22" long piece of 3/4" redwood 3/4" wide and then crosscut four pieces 5" long for the trim (**I**). Lay out the four trim porches (**J**) on a piece of 3/4" redwood as shown in the trim porch detail drawing. Cut the pieces on a band saw and sand the edges smooth. Glue and nail the pieces to the house at floor level. Glue and nail the pieces to the back wall. Lay out the centers for the screws that will secure the back wall to the house. Drill and countersink the holes. Put the rear wall in place and add the six flat head wood screws (**K**).

Use a non-toxic finish for the birdhouse. Do not paint the interior of the house nor the inside of the entrance holes. We used exterior latex enamel to paint our birdhouse. Paint the birdhouse your favorite color and the roof a contrasting color. We painted the exterior walls, porches and trim a light brown and the top a light green. The

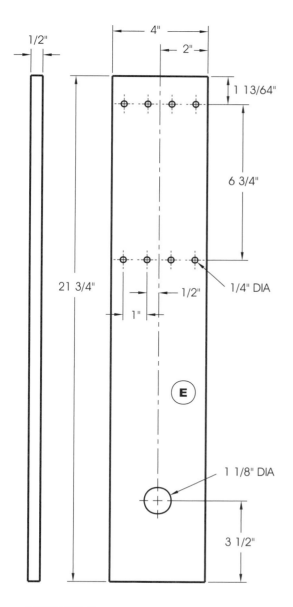

LEFT SIDE DETAIL

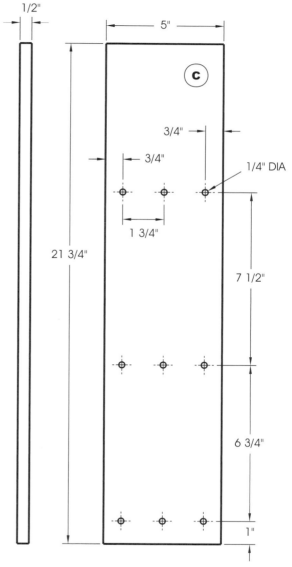

BACK DETAIL

LIST OF MATERIALS

	DESCRIPTION	MATERIAL	FINISHED SIZE
A	FLOOR (3)	REDWOOD	3/4 x 4 x 4
B	FRONT	PLYWOOD	1/2 x 5 x 21 3/4
C	BACK	PLYWOOD	1/2 x 5 x 21 3/4
D	RIGHT SIDE	PLYWOOD	1/2 x 4 x 21 3/4
E	LEFT SIDE	PLYWOOD	1/2 x 4 x 21 3/4
F	TOP	REDWOOD	3/4 x 6 1/2 x 8 1/2
G	BOTTOM	REDWOOD	3/4 x 6 1/2 x 8 1/2
H	CEILING	REDWOOD	3/4 X 4 X 4
I	TRIM (4)	REDWOOD	3/4 x 3/4 x 5
J	TRIM PORCH (4)	REDWOOD	3/4 x 1 3/4 x 6 1/2
K	SCREW (6)	BRASS	FL HD No 8 X 1 1/2
L	EYEBOLT	STEEL	3/16 x 2 GALVANIZED
M	WASHER	STEEL	No 10 GALVANIZED
N	NUT	STEEL	10-32 GALVANIZED
O	BOX NAIL (40)	STEEL	4d GALVANIZED

bottom and the underside of the top were left unpainted and will weather a warm silver brownish color over time.

Insert the eyebolt (**L**) into the hole in the top and add the washer (**M**) and nut (**N**). Put a dab of glue on the threads to keep the nut from coming loose.

Hang the birdhouse tower in a tree in your yard and wait for the birds to come calling. If you put out a bird feeder before you hang the house, several birds should be in the area ready for the birdhouse. **tp**

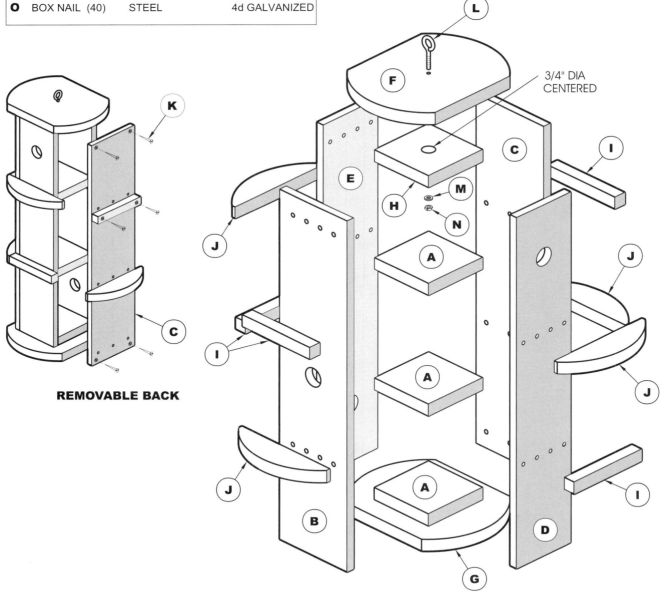

CAROLINA WREN TOWER BIRDHOUSE ASSEMBLY

Bluebird Birdhouse

The bluebird birdhouse is a standard design with a removable sloping roof. The roof is made removable for easy cleaning of the house. The birdhouse is made of 1/2" construction grade plywood with the rough side to the inside except for the roof which has the rough side to the outside for a contrasting texture. It is best to have the rough side inside so the birds can cling to the inside wall easier.

Rip a 28" long piece of 1/2" thick plywood 5" wide for the floor (**A**) and the two sides (**B**). Crosscut a piece 5" long for the floor. Lay out the centers of the four drain holes 3/8" from the edges and drill with a 1/4" diameter bit.

Lay out the two sides on the remaining piece of 5" wide plywood. Measure up 9" from the bottom edge for the front and 11" for the back. Cut the two pieces on a band saw and sand the edges smooth.

Rip a 20" long piece of 1/2" thick plywood 6" wide for the front (**C**) and back (**D**). Set your table saw blade to the same angle as the roof slope on the side and cut the front with the long 9" dimension on the inside and the bevel going down to the outside face. Cut the back with the 11" dimension on the inside and the bevel going up to the outside face. Lay out the centers for

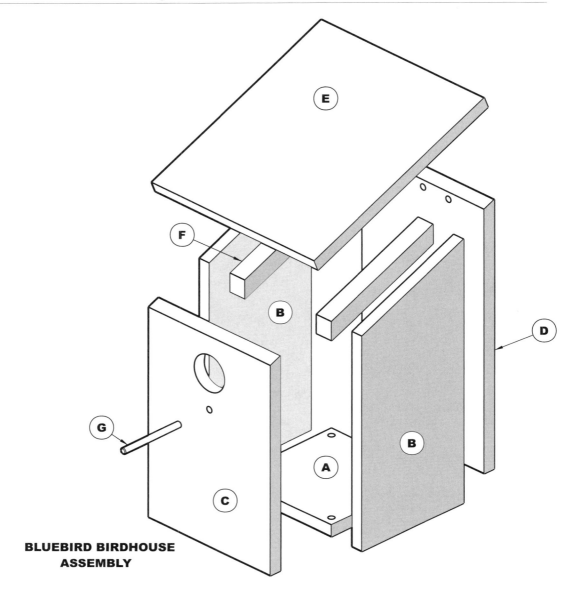

BLUEBIRD BIRDHOUSE ASSEMBLY

LIST OF MATERIALS

	DESCRIPTION	MATERIAL	FINISHED SIZE
A	FLOOR	PLYWOOD	1/2 x 5 x 5
B	SIDE (2)	PLYWOOD	1/2 x 5 x 11
C	FRONT	PLYWOOD	1/2 x 6 x 9
D	BACK	PLYWOOD	1/2 x 6 x 11 3/16
E	ROOF	PLYWOOD	1/2 x 8 x 9
F	SUPPORT (2)	PINE	3/4 x 3/4 x 5 1/4
G	PERCH	DOWEL	1/4 DIA x 2 3/4
H	BOX NAIL (6)	STEEL	4d GALVANIZED

the entrance hole and perch hole in the front wall, and the three ventilation holes in the back wall. Drill the 1 1/2" diameter entrance hole in the front and the 1/4" diameter perch hole. Drill the three 1/4" diameter ventilation holes in the back.

Glue and nail, using the 4 penny box nails (**H**), the sides to the edge of the floor. Then glue and nail the front to the edge of the floor and the edge of the sides. Next, glue and nail the back to the edge of the floor and the edge of the sides.

Lay out the roof (**E**) 8" wide by 9" long on a piece of 1/2" plywood. Cut the pieces on

Bluebird Birdhouse

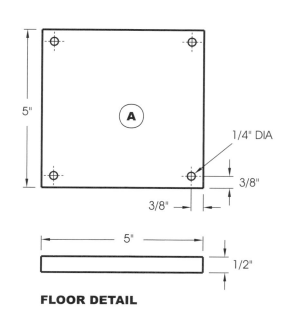

FLOOR DETAIL

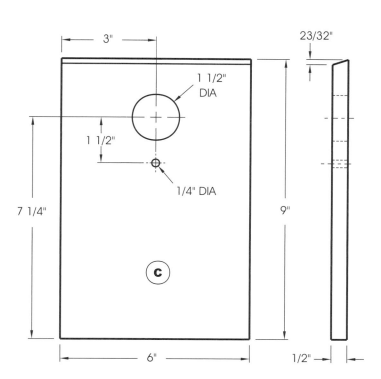

FRONT DETAIL

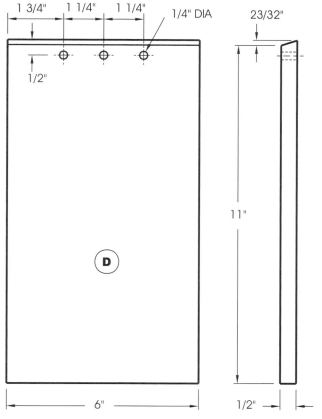

BACK DETAIL

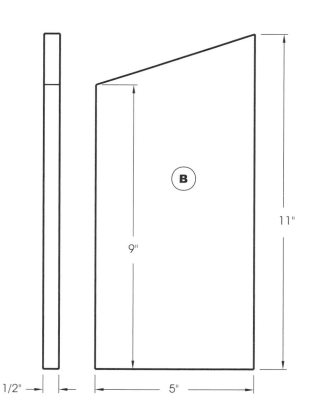

SIDE DETAIL (MAKE 2)

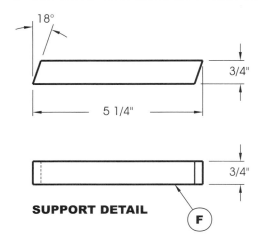

SUPPORT DETAIL

a band saw and sand the edges smooth.

Rip a 12" long piece of 3/4" thick pine 3/4" wide for the two supports (**F**). Cut the two supports 5 1/4" long and sand the 18 degree angles to both ends. Mark the underside of the roof for the supports. Glue and nail the supports to the roof.

Use a non-toxic finish for the birdhouse. Do not paint the interior of the house nor the inside of the entrance hole. We used exterior latex enamel to paint our birdhouse. Paint the birdhouse your favorite color and the roof a contrasting color. We painted the exterior walls and the exposed underside of the roof a light brown and the top of the roof a light green.

When the paint dries glue the perch (**G**) into the hole in the front.

Attach a pipe flange to the bottom of the floor with some wood screws and mount the birdhouse on a pipe five foot from the ground. Secure the post in the ground with some method of getting the birdhouse to the ground level for cleaning. A wooden mount and pole could also be used. If you put out a bird feeder before you erect the house, several birds should be in the area ready for the birdhouse. **tp**

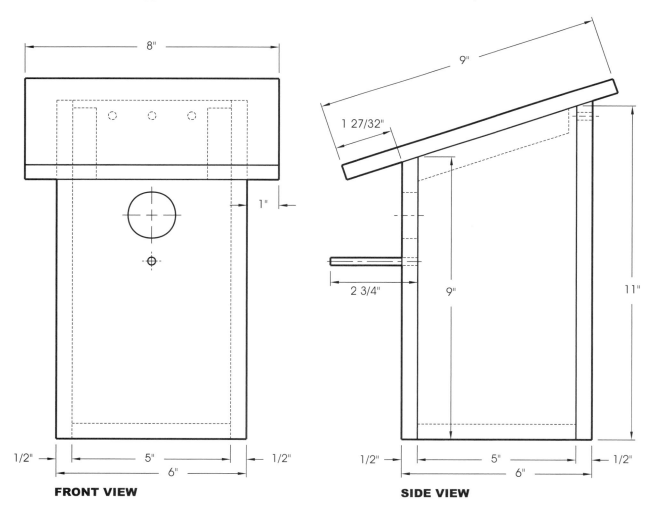

FRONT VIEW **SIDE VIEW**

Western Martin Triplex Birdhouse

The purple martin is one of few birds that will nest in groups. Some of the birdhouses that have been built for these birds are very big with up to 10 nesting boxes in one house. However, to date the western martin will not nest in such large groups; they will nest in groups of three. Perhaps as areas of the west get more congested the birds will nest in larger groups. Our birdhouse for the western martin will house three families of birds and is securly fastened to the top of a pole. The house is 10 inches by 20 inches with a porch for each opening. The entrance holes are on three different sides to give the birds some privacy. The roof is removable for easy cleaning of all three homes. The birdhouse is made of 1/2" construction grade plywood with the rough side in except for the roof. The rough side out for the roof gives a nice contrast to the slightly smoother sides. Other pieces are made of 5/8" western red cedar.

Rip two 20" long pieces of 1/2" thick plywood 9" wide for the floor (**A**) and ceiling (**B**). Then rip two 20" long pieces of 1/2" thick plywood 7" wide for the front (**C**) and back (**D**). Cut the four pieces to a length of 19". Use a stop block attached to the tabletop to insure the three pieces are the same length.

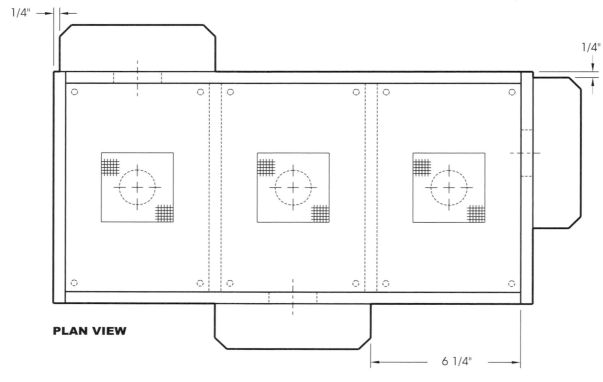

PLAN VIEW

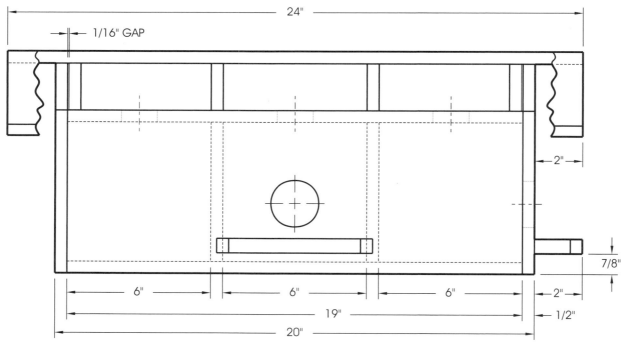

FRONT VIEW

Lay out the centers for the twelve drain holes in the floor. Drill the holes with a 1/4" diameter drill bit.

Lay out the centers for the three vent holes in the ceiling. Drill the holes with a 1 1/2" diameter hole saw.

Lay out the centers for the entrance holes in the front and back walls. Drill the holes with a 2" diameter hole saw. Sand the edges with a 1" diameter drum sander and break the front and back edges with a piece of sandpaper.

Lay out the right side (**E**) on a piece of 1/2" plywood and cut out. Lay out the center for the entrance hole and the ventilation hole in the side wall. Drill the entrance hole with a 2" diameter hole saw and the vent hole with a 1" diameter hole saw.

Lay out the left side (**F**) on a piece of 1/2" plywood and cut out. Lay out the center for the vent hole in the side wall. Drill the vent hole with a 1" diameter hole saw.

Rip a piece of 20" long 5/8" thick red cedar 2" wide and then crosscut into three pieces 6 1/2" long for the porches (**G**). Lay out the 1/2" by 1/2" chamfers on the two front corners on each of the three pieces. Cut the chamfer and then sand smooth.

Lay out the position of the porches on the front wall, back wall and the right side wall. Glue and nail, using the 4 penny box nails (**N**), the porches to the walls from the inside.

Rip a piece of 19" long piece of 1/2" thick plywood 6" wide for the two dividers (**H**). Cut the pieces into two pieces 9" long. Use a stop block attached to the tabletop to insure the two pieces are the same length.

Glue and nail the front wall to the edge of the floor. Then glue and nail the edges of the two dividers to the floor and the front wall. Next, glue and nail the back wall to the edge of the floor and the edge of the two dividers. Then glue and nail the right side to the edge of the floors and the edge of the front and back. And finally glue and nail the left side to the edge of the floor and the edge of the front and back.

Rip two 25" long pieces of 1/2" thick plywood 8 1/2" wide for the two roofs (**I**). Cut the two pieces to a length of 24". Use a stop block attached to the tabletop to insure the two pieces are the same length. Now set your table saw blade to 30 degrees and set the fence to give a 7 3/4" dimension to the top of the roof. Cut a scrap piece to check the setup and make any adjustments necessary. When the setup is correct cut the two roof pieces. Place the two roof sections on the peaks of the side walls for a test fit. When everything looks right put glue on both beveled edges and put them together on the birdhouse. Add short strips of masking tape across the peak to keep the roof together.

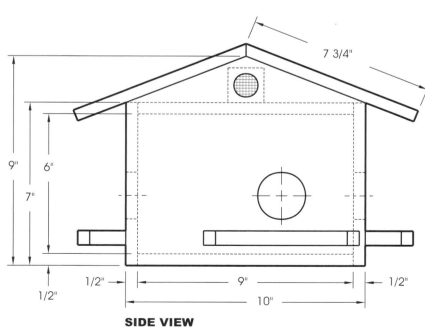

SIDE VIEW

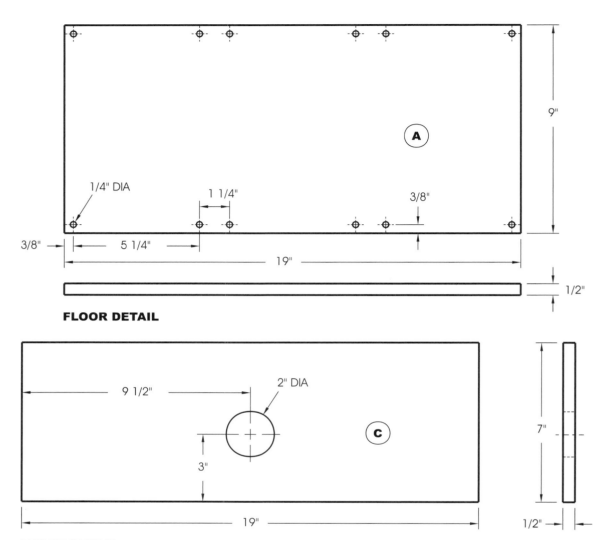

FLOOR DETAIL

FRONT DETAIL

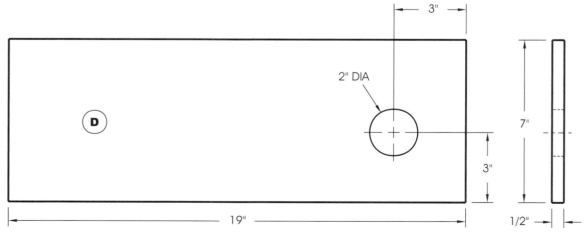

BACK DETAIL

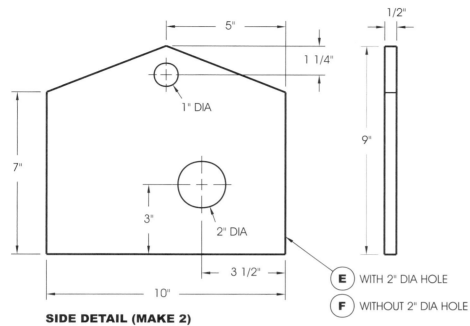

SIDE DETAIL (MAKE 2)

E WITH 2" DIA HOLE
F WITHOUT 2" DIA HOLE

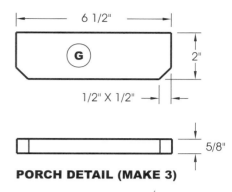

PORCH DETAIL (MAKE 3)

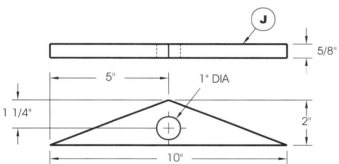

ROOF SUPPORT DETAIL (MAKE 4)

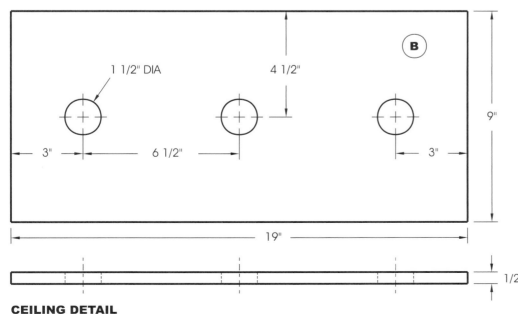

CEILING DETAIL

While the glue dries lay out the four roof supports (**J**) on a piece of 5/8" red cedar. Mark the centers for the vent holes and drill with a 1" diameter hole saw. Use a band saw to cut the pieces away from the line and then sand to the line. Remove the roof from the birdhouse and mark the location of the four supports on the underside of the roof. Glue the supports to the underside of the roof. When dry, remove the small strips of masking tape and put one long piece of cloth tape along the peak of the roof for the roof cover (**K**).

Use a non-toxic finish for the birdhouse. Do not paint the interior of the house nor the inside of the entrance holes. We used exterior latex enamel to paint our birdhouse. Paint the birdhouse your favorite color and the roof a contrasting color. Martins prefer white walls so we painted the exterior walls white and the roof a light green. The remaining porches were left unpainted and will weather to a warm brownish gray color over time.

Staple the three screens (**L**) over

26 Simple Birdhouses and Feeders of Wood

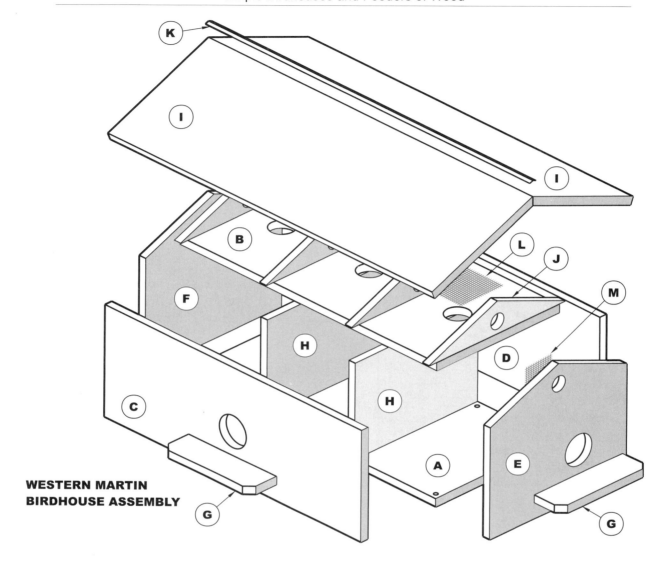

WESTERN MARTIN BIRDHOUSE ASSEMBLY

	LIST OF MATERIALS		
	DESCRIPTION	MATERIAL	FINISHED SIZE
A	FLOOR	PLYWOOD	1/2 x 9 x 19
B	CEILING	PLYWOOD	1/2 x 9 x 19
C	FRONT	PLYWOOD	1/2 x 7 x 19
D	BACK	PLYWOOD	1/2 x 7 x 19
E	RIGHT SIDE	PLYWOOD	1/2 x 9 x 10
F	LEFT SIDE	PLYWOOD	1/2 x 9 x 10
G	PORCH (3)	RED CEDAR	5/8 x 2 x 6 1/2
H	DIVIDER (2)	PLYWOOD	1/2 x 6 x 9
I	ROOF (2)	PLYWOOD	1/2 x 7 3/4 x 24
J	ROOF SUPPORT (4)	RED CEDAR	5/8 x 2 x 10
K	ROOF COVER	CLOTH TAPE	1 1/2 WIDE x 24
L	SCREEN (3)	HDW CLOTH	1/8 MESH 3 x 3
M	SCREEN (2)	HDW CLOTH	1/8 MESH 1 1/2 x 2
N	BOX NAIL (40)	STEEL	4d GALVANIZED

the holes in the top of the ceiling. The hardware cloth to use is the 1/8" mesh galvanized metal material not the ordinary fly screen used on most screen windows now a days. Staple the two screens (**M**) inside over the holes in the right side and left side walls.

Attach some type of fitting to the bottom of the birdhouse and mount to a pole 5 feet above the ground. The employment of the pole should allow for some method of getting the birdhouse to ground level for cleaning. See the chapter on Mounting and Poles, page 58 for some ideas. **tp**

Chickadee Birdhouse

This birdhouse does not look like the common design birdhouse and is well suited for the chickadee birds. The floor is the common 4 by 4 inches, but the house is quite deep, which makes this house appear much taller than the common birdhouse. This birdhouse could be hung from a branch in a tree, but it is more suited for a pole mounted birdhouse. This house is constructed exclusively of 1/2" thick construction grade plywood except for the perch and the roof support.

Rip a 24" long piece of 1/2" thick plywood 4" wide. Then crosscut a piece 4" long for the floor (**A**). Lay out the centers for the four drain holes in the four corners. Drill the holes with a 1/4" drill bit.

Next rip a 22" long piece of 1/2" thick plywood 5" wide. Then crosscut the piece in the middle and use one piece for the front (**B**) and the other for the back (**C**). Lay out the peak and the centers for the three holes in the front using the dimensions from the front detail drawing. Cut the peak on the band saw, cutting away from the line and then sand to the line. Drill the 1 1/8" diameter hole for the entrance, the 1/4" diameter hole for the perch and the 3/16" diameter hole for the screw that will attach the roof to the house. Lay out the peak and the centers for the six holes in the back using the dimensions from the back detail drawing. Cut the peak on the band saw, cutting away from the line and then sand to the line. Drill the five 1/4" diameter holes for ventilation and the 3/16" diameter hole for the other screw that will attach the roof to the house.

Take the remaining piece of 4" wide plywood that was cut earlier and use this for the two sides (**D**). Crosscut the piece in half and then set the table saw blade to the same angle as the front wall peak. Cut a piece of scrap 1/2" thick plywood and check the piece against the roof angle of the front wall. Make adjustments to the angle of the blade and then run both of the 4" wide pieces through the saw. Adjust the table saw blade back to 90 degrees to the table and cut the bottom of the side wall edges to length.

Next rip a 9 1/2" long piece of 1/2" thick plywood 7 1/4" wide for the two roofs (**E**) and then crosscut into two equal pieces. Cut the bevel on the top edge of the two roofs to match the front wall peak.

Next we will cut the bevel on the top edge of the roof support (**F**) before it is cut to size. Set the table saw blade to the same angle as the front wall peak. Set the fence and take an 8" long 3 1/2" wide 3/4" thick red cedar and cut one side of the bevel. Flip the piece of wood over and cut the second bevel. This method will guarantee that the top of the bevel is in the center of the 3/4" thickness. Adjust the table saw blade back to 90 degrees to the table and cut the piece 3/4" from the top of the peak. Crosscut the piece to the finished 4" long dimension.

Test fit the sides to the front and back around the floor, and make any adjustments necessary. When everything fits properly glue and nail, using the 4 penny box nails (**I**), the sides to the edge of the floor. Then glue and

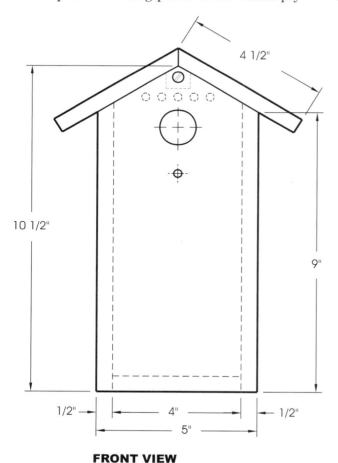

FRONT VIEW **SIDE VIEW**

Chickadee Birdhouse

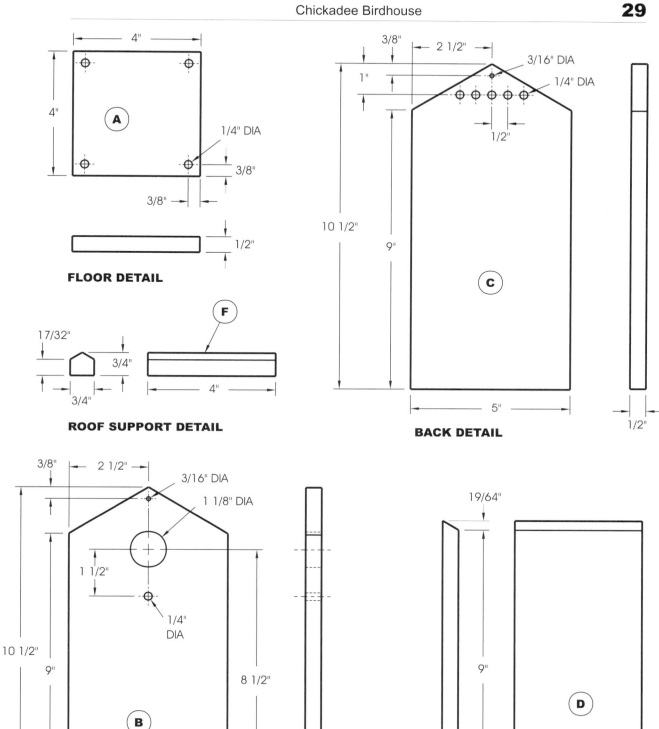

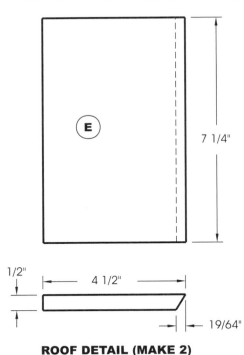

ROOF DETAIL (MAKE 2)

entering the roof where the two halves join.

Use a non-toxic finish for the birdhouse. Do not paint the interior of the house nor the inside of the entrance hole. We used exterior latex enamel to paint our birdhouse. Paint the birdhouse your favorite color and the roof a contrasting color. We painted the exterior walls and the exposed underside of the roof a light brown color and the roof and roof edges a light green. Apply two or three coats and allow to dry overnight between each coat.

Put the roof on the birdhouse and add the two round head wood screws (**J**). Glue the perch (**G**) into the hole in the front wall.

Mount your chickadee birdhouse on the top of a 10 to 15 foot pole in your yard. It will not be long before you have your first tenants with some young ones to follow. **tp**

nail the front and back to the edge of the floor and to the edges of the sides. Use waterproof glue for all joints. Test the fit of the two roof pieces to the house and make any adjustments. Apply glue to the ridge bevel of the two roof sections and place both pieces on the top of the birdhouse. Only the ridge bevel gets glued since the roof needs to be removable for cleaning. Put two strips of masking tape across the roof pieces and allow to dry. Glue the ridge beam to the underside of the roof with equal clearance to the front and back.

The roof cover (**H**) is a strip of 1 1/2" wide cloth tape. Put half of the width of the tape on one side of the roof at the peak and fold it down to other side. This will help to keep the rain from

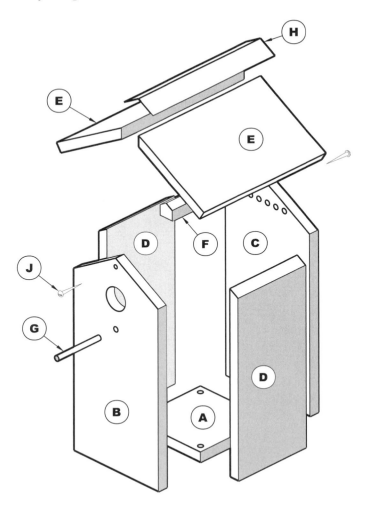

CHICKADEE BIRDHOUSE ASSEMBLY

LIST OF MATERIALS

	DESCRIPTION	MATERIAL	FINISHED SIZE
A	FLOOR	PLYWOOD	1/2 x 4 x 4
B	FRONT	PLYWOOD	1/2 x 5 x 10 1/2
C	BACK	PLYWOOD	1/2 x 5 x 10 1/2
D	SIDE (2)	PLYWOOD	1/2 x 4 x 9 3/8
E	ROOF (2)	PLYWOOD	1/2 x 4 1/2 x 7 1/4
F	ROOF SUPPORT	RED CEDAR	3/4 x 3/4 x 4
G	PERCH	DOWEL	1/4 DIA x 2 1/2
H	ROOF COVER	CLOTH TAPE	1 1/2 WIDE x 7 1/4
I	BOX NAIL (20)	STEEL	4d GALVANIZED
J	WOOD SCREW (2)	BRASS	1 1/4 x No 8 RD HEAD

Small Hanging Bird Feeder

This hanging bird feeder is small and was built to feed small birds like sparrows, finches and wrens. These small birds do like this feeder and its open design allows several to feed at the same time. However, the small size does not stop scrub jays, doves and an occasional mountain dove from using the feeder. This feeder is built of 3/8" thick construction grade plywood and redwood.

Lay out the floor (**A**) 6 1/2" by 8 1/2" on 3/8" thick plywood. Measure in 1/4" from the edges in each corner for the drain hole center lines. Cut the floor to size and drill the four 1/4" diameter holes. Sand the edges smooth.

Next lay out the upright (**B**) using the upright detail drawing on a 3/4" thick piece of redwood. Using a band saw cut away from the lines and then sand to the lines. Lay out the outline of the upright base on the underside of the floor. Drill two No. 8 countersunk screw holes in the floor. Transfer the dimensions of one of the holes to the upright base and drill a pilot hole into the upright. Attach the upright to the base using the 1 1/4" long No. 8 flat head wood screw (**K**) and position it over the other hole. Drill the pilot hole in the upright using the other hole in the base. Remove the wood screw and set the

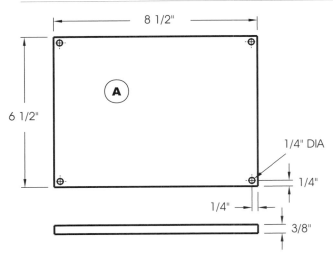
FLOOR DETAIL

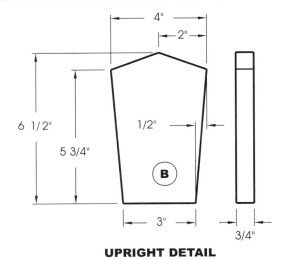
UPRIGHT DETAIL

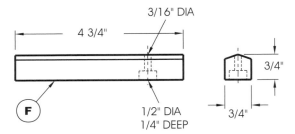
ROOF SUPPORT DETAIL (MAKE 2)

upright aside for now.

Cut the two side edgings (**C**) and the two front edgings (**D**) 1" wide from a 10" long piece of 1/2" thick redwood. Cut the side edging pieces to a length of 6 1/2" and drill pilot holes for the 4 penny box nails (**L**). Glue and nail the side edging pieces to the edge of the floor. Position the front edging pieces against the floor and mark the ends. Cut to length and drill pilot holes in these pieces. Glue and nail the front edging pieces to the edge of the floor.

Next lay out the two roofs (**E**) 5" wide by 10 1/4" long on a piece of 3/8" thick plywood. Use the upright part and set the table saw blade to the angle of the roof angle. Position the fence on the side with the blade tip facing the fence. Measure along the table and set the fence 4 1/2" from the blade. Cut the two roof sections

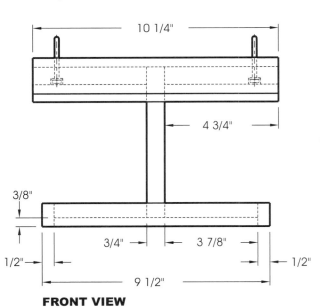
FRONT VIEW

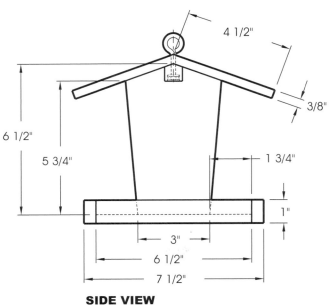
SIDE VIEW

Small Hanging Bird Feeder

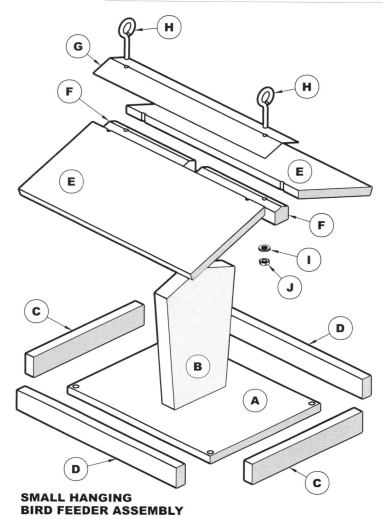

SMALL HANGING BIRD FEEDER ASSEMBLY

LIST OF MATERIALS

	Description	Material	Finished Size
A	FLOOR	PLYWOOD	3/8 x 6 1/2 x 8 1/2
B	UPRIGHT	REDWOOD	3/4 x 4 x 6 1/2
C	SIDE EDGING (2)	REDWOOD	1/2 x 1 x 6 1/2
D	FRONT EDGING (2)	REDWOOD	1/2 x 1 x 9 1/2
E	ROOF (2)	PLYWOOD	3/8 x 4 1/2 x 10 1/4
F	ROOF SUPPORT (2)	REDWOOD	3/4 x 3/4 x 4 3/4
G	ROOF COVERING	CLOTH TAPE	1 1/2 x 10 1/4
H	EYEBOLT (2)	STEEL	3/16 x 2 GALV.
I	WASHER (2)	STEEL	NO 10 GALV.
J	NUT (2)	STEEL	10-32 GALV.
K	WOOD SCREW (4)	STEEL	1 1/4 x No 8 FL HD
L	BOX NAIL (10)	STEEL	4d GALVANIZED

to width with the outside (rough side) down against the table.

Lay the two roof sections on a flat surface, outside up and with the top edge of each bevel against the other and put two strips of masking tape across the peak. Mark the top and underside of the roof with a line parallel to and 4 3/4" from the side edge. Clamp the upright in a vise and place the roof section on the peak of the upright and line up the line on the underside of the roof with the side of the upright. Drill a No. 8 countersunk screw hole through the roof and into the upright 3/8" from the line. Attach the roof to the upright with a 1 1/4" long No. 8 flat head screw. Drill another countersunk screw hole through the other roof section and into the upright. Remove the first screw and add glue to the roof bevel and the upright edge. Put the roof on the upright and add both wood screws.

With the table saw still set to the roof angle of the upright cut a 12" long piece of 3/4" thick redwood for the roof support (**F**). Move the fence close enough to the blade so it will cut a complete bevel. Turn the piece of wood over and cut another bevel. This will give an equal bevel on both sides of the 3/4" thick piece of wood. Put the blade back to 90 degrees to the table and cut the roof support 3/4" tall. Cross cut to two pieces 4 3/4" long. Drill the 3/16" diameter hole through the support and drill the 1/2" diameter counterbore 1/4" deep. Remove the upright from the vise and glue both roof supports to the underside of the roof and the upright. Drill the 3/16" diameter through the roof using the hole in the supports as a guide.

Add glue to the bottom of the upright and join it to the floor with the two wood screws.

Add the roof covering (**G**) to the peak of the roof. With an awl poke a hole in the tape at both holes in the roof.

We painted our feeder, roof eves a light brown color and the roof a light green. Allow to dry overnight and then add the eyebolt (**H**), washer (**I**) and nut (**J**). Put a dab of white glue on the threads to keep the nut from coming loose. This is not a permanent attachment; the nut can still be removed if needed.

Hang the small hanging feeder from a branch of a tree and soon you will have birds visiting your new feeder. Be prepared to fill this small feeder with birdseed once or twice a day.

tp

Hopper Wall Mounted Bird Feeder

This hopper wall mounted bird feeder will hold a large amount of bird seed and dispense just a limited amount to the large platform below the opening at the bottom. It is best hung high up on the side of a building or nailed to the side of a post.

This feeder is built exclusively of 1/2" thick construction grade plywood. The floor extends out a good distance from the sloping front and the roof also extends out to give the floor protection from the rain.

Hopper Wall Mounted Bird Feeder

Lay out the floor (**A**) 6" wide by 8 1/2" long on a piece of 1/2" thick plywood. Measure in 3/8" from the edges in the two forward corners for the drain hole center lines. Cut the floor to size and drill the two 1/4" diameter holes. Sand the edges smooth.

Lay out the back (**B**) 6" wide by 15 1/2" long on a piece of 1/2" thick plywood. We will cut the top to the finished length once we have established the roof slope of the side walls.

Now lay out the two sides (**C**) using the side detail drawing on two 6" wide by 14" long pieces of 1/2" thick plywood. Using a band saw cut away from the lines and then sand to the lines.

Next lay out the front (**D**) 6" wide by 13" long on a piece of 1/2" thick plywood. Set the table saw blade to match the angle at the top front edge of the side wall. Place the front edge of the side wall on the table. Use a scrap of plywood and do a trial cut. Check the cut against the side wall and make any adjustment to the blade angle. Cut trial cuts until the right angle is made and then cut the front wall to a length of 12 9/16" to the outside edge.

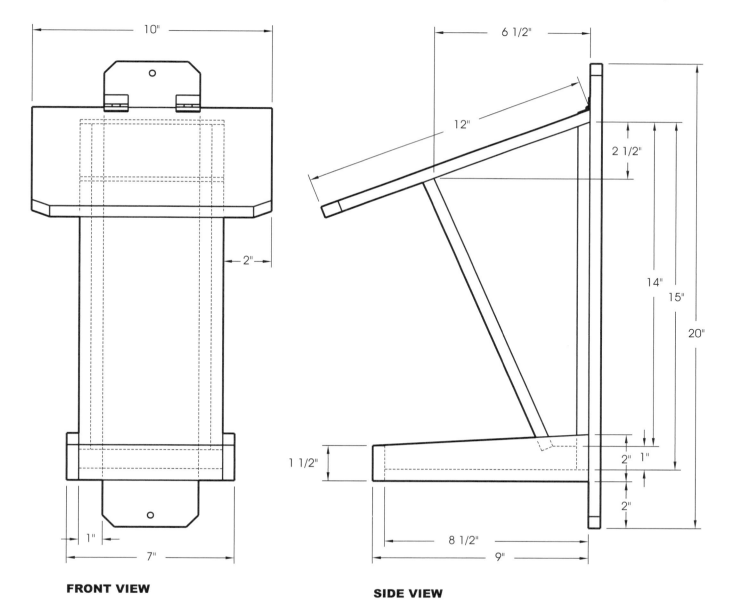

FRONT VIEW **SIDE VIEW**

LIST OF MATERIALS

	DESCRIPTION	MATERIAL	FINISHED SIZE
A	FLOOR	PLYWOOD	1/2 x 6 x 8 1/2
B	BACK	PLYWOOD	1/2 x 6 x 15
C	SIDE (2)	PLYWOOD	1/2 x 6 x 13 13/16
D	FRONT	PLYWOOD	1/2 x 6 x 12 9/16
E	ROOF	PLYWOOD	1/2 x 10 x 12
F	BACK SUPPORT	PLYWOOD	1/2 x 4 x 20
G	FRONT EDGING	PLYWOOD	1/2 x 1 1/2 x 6
H	SIDE EDGING (2)	PLYWOOD	1/2 x 2 x 9
I	HINGE (1 PAIR)	BRASS	1 x 1
J	BOX NAIL (30)	STEEL	4d GALVANIZED

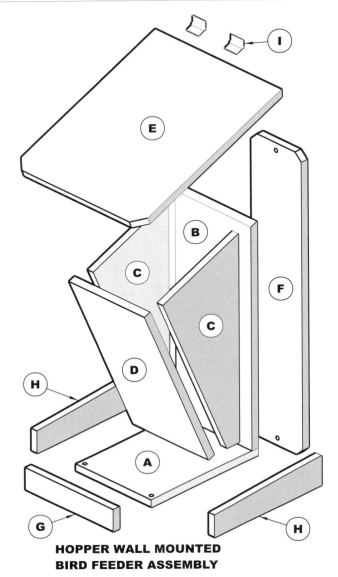

HOPPER WALL MOUNTED BIRD FEEDER ASSEMBLY

Now adjust the table saw blade to match the angle at the top back corner of the side wall. Place the back edge of the side wall on the table. Use a scrap of plywood and do a trial cut. Do trial cuts in a scrap piece of plywood and then cut the back to a length of 15" to the outside long edge.

Glue and nail, using the 4 penny box nails (**J**), the two side wall edges to the back 1" above the bottom edge of the back and flush with the top edge. Glue and nail the front to the edges of the side walls and then glue and nail the floor to the bottom edge of the back. Keep the drain holes to the front.

Lay out the roof (**E**) 10" wide by 12 1/2" long on a piece of 1/2" thick plywood. This is 1/2" wider than the finished size for the bevel at the top. Set the table saw blade to the same angle as the side wall roof angle. Take trial cuts to a scrap piece of plywood and then cut the back to a length of 12" to the outside edge. Lay out the two chamfers at the front two corner of the roof 3/4" by 3/4". Cut the chamfers and sand all the edges.

Next lay out the back support (**F**) 4" wide by 20" long on a piece of 1/2" thick plywood. Lay out the 1/2" by 1/2" chamfers and the center lines for the two mounting holes 2" from the edge and 1/2" from both ends. Cut the chamfers and drill the 1/4" diameter holes. Glue and nail the back support to the back with the lower edge of the back support 2" below the bottom of the floor and 1" from the side.

Cut the front edging (**G**) 1 1/2" wide by 6" long from a piece of 1/2" thick plywood. Glue and nail the front edging pieces to the front edge of the floor.

Lay out the two side edgings (**H**) 2" wide at the back and 1 1/2" wide at the front by 9" long from a piece of 1/2" thick plywood. Cut the two edgings and sand the edges.

Hopper Wall Mounted Bird Feeder 37

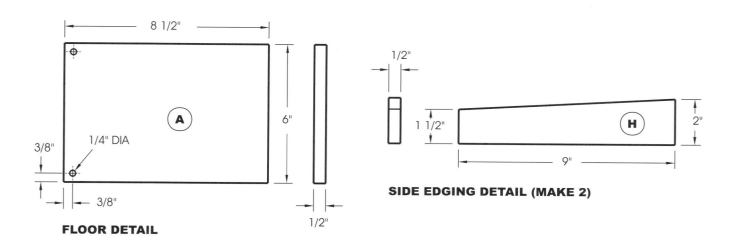

FLOOR DETAIL

SIDE EDGING DETAIL (MAKE 2)

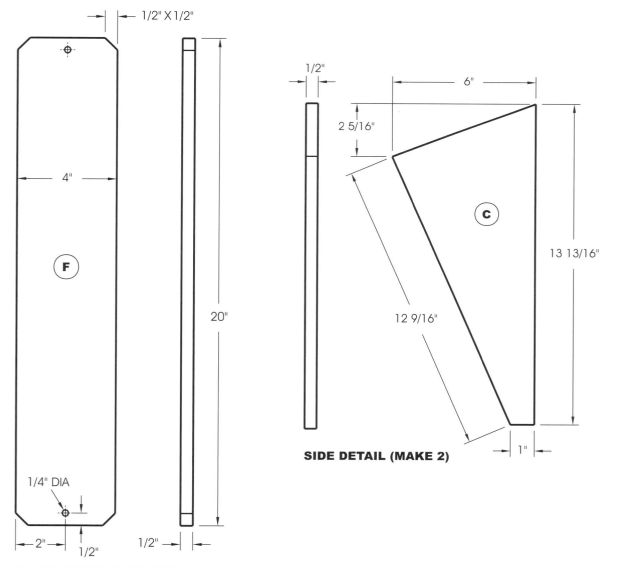

BACK SUPPORT DETAIL

SIDE DETAIL (MAKE 2)

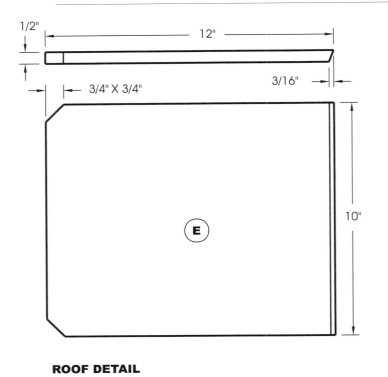

ROOF DETAIL

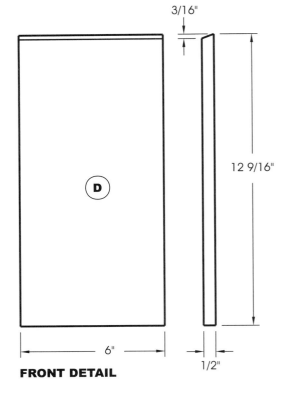

FRONT DETAIL

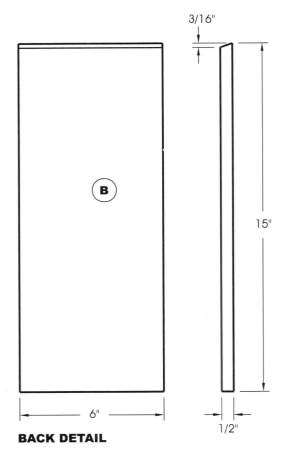

BACK DETAIL

Glue and nail the side edging pieces to the edge of the floor, the edge of the front edging and the edge of the back.

Use a non-toxic finish for the bird feeder and do not paint the inside surface of the edging or the floor of the feeder. We used an exterior latex enamel to paint our feeder. Paint the feeder your favorite color and the roof a contrasting color. Paint the bird feeder, outside edge of the trim and the exposed underside of the roof a light color and the roof and the roof edges a darker color. We painted our bird feeder a light tan with a light green roof. Apply two or three coats and allow to dry.

When dry add the pair of hinges (I) to the roof and back support. With a point of an awl pierce the plywood at the hinge screw holes. Now drill small pilots holes at each of the marks and screw the hinges to the plywood roof and support back.

Nail the hopper wall mounted feeder to the side of a barn or garage and soon you will have birds visiting your feeder. **tp**

Covered Hopper Bird Feeder

This covered hopper bird feeder will hold a large amount of bird seed. The amount of seed in the feeder can be seen through the two glass sloping walls. The glass does not go down to the floor which allows a controlled amount of seed to spill out on to the floor. This control along with the small floor area saves seed as the birds can not get into the floor and thrash around and spill the seed to the ground. This feeder is built of 1/2" thick construction grade plywood and western red cedar.

Lay out the floor (**A**) 5" wide by 10" long on 1/2" thick plywood. Measure in 3/8" from the edges in each corner for the drain hole centerlines. Cut the floor to size and drill the four 1/4" diameter holes. Sand the edges smooth.

Lay out the two sides (**B**) using the side detail drawing on two 6" wide by 8 1/2" long pieces of 1/2" thick plywood. Using a band saw cut away from the lines and then sand to the lines. Cut the two 1/8" slots 1/4" deep 5/16" from the edges for the glass. Glue and nail, using the 4 penny box nails (**O**), the floor to the bottom edge of the two side walls 1 3/8" from the front edge and flush with the side edge of the floor.

Cut the two front edgings (**C**) 1 1/2" wide by 10" long from a piece of 5/8" thick west-

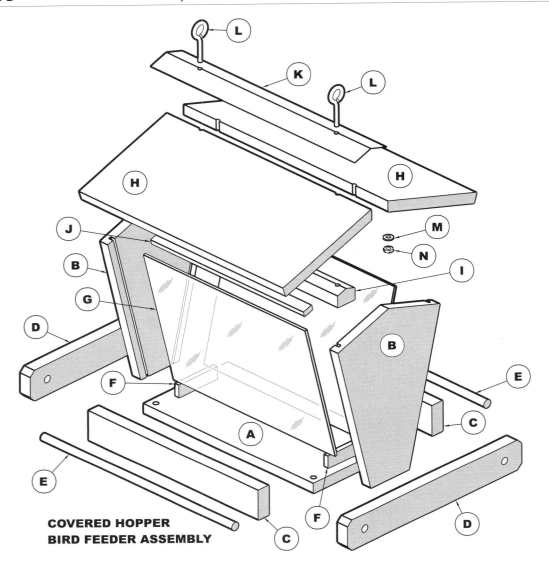

COVERED HOPPER BIRD FEEDER ASSEMBLY

ern red cedar. Glue and nail the front edging piece to the edge of the floor. First drill pilot holes in the cedar to prevent splitting.

Cut the two perch supports (**D**) 1 1/2" wide by 10 1/4" long from a piece of 5/8" thick western red cedar. Lay out the center of the perch holes and the 1/4" by 45 degrees chamfers at each corner as shown in the perch support detail drawing. Drill the 3/8" diameter holes and then cut the chamfers and sand. Glue and nail the perch support pieces to the edge of the floor and the front edging

	LIST OF MATERIALS		
	DESCRIPTION	MATERIAL	FINISHED SIZE
A	FLOOR	PLYWOOD	1/2 x 5 x 10
B	SIDE (2)	PLYWOOD	1/2 x 6 x 8 1/2
C	FRONT EDGING (2)	RED CEDAR	5/8 x 1 1/2 x 10
D	PERCH SUPPORT (2)	RED CEDAR	5/8 x 1 1/2 x 10 1/4
E	PERCH (2)	DOWEL	3/8 DIA x 11 1/4
F	GLASS STOP BLOCK (2)	RED CEDAR	5/8 x 1/4 x 2 1/4
G	FRONT (2)	GLASS	1/8 x 7 1/8 x 9 1/2
H	ROOF (2)	PLYWOOD	1/2 x 5 1/2 x 12
I	ROOF SUPPORT	RED CEDAR	1 X 3/4 x 9
J	ROOF STOP	RED CEDAR	1/4 x 1/2 x 9
K	ROOF FLASHING	GALV. TIN	2 WIDE x 12
L	EYEBOLT (2)	STEEL	3/16 x 2 GALV.
M	WASHER (2)	STEEL	No 10 GALV.
N	NUT (2)	STEEL	10-32 GALV.
O	BOX NAILS (14)	STEEL	4d GALVANIZED
P	WOOD SCREW (2)	BRASS	1/2 x No 6 FL HD

ends with the front edge 2" forward of the front edging. Again first drill pilot holes in the cedar to prevent splitting.

Cut the two perches (**E**) to 11 1/4" long from a piece of 3/8" diameter dowel. Slide the dowel into the holes in the perch supports for a test fit. Slide the dowel out about 1" and put a dab of glue on the round part at each end and slide back into the holes.

Cut two glass stop blocks (**F**) 1/4" wide by 2 1/4" long from a 5/8" thick piece of western red cedar. Glue the two pieces to the bottom of the floor and the inside of the side walls.

Cut the 1/8" piece of glass into two pieces each 7 1/8" wide by 9 1/2" long using a glass cutter. It may be best to have these pieces cut to finished size by a glass or picture frame company. Glass is not the easiest material to cut. Slide the two glass fronts (**G**) into the slots in the side walls.

Lay out the two roofs (**H**) 6" wide by 12" long on a piece of 1/2" thick plywood. This is 1/2" wider than the finished size for the bevel at the top. Set the table saw blade to the same angle as the side wall roof angle. Set the fence 5 1/2" from the blade as measured along the table. Cut both roof pieces to width.

With the table saw blade set to the angle of the roof cut a 10" long piece of 1" thick western red cedar for the roof support (**I**). Move the fence close enough so the blade will cut a complete bevel. Turn the piece of wood over and cut another bevel. This will give an equal bevel on both sides of the 1" thick piece of wood. Put the blade back to the 90 degree cutting angle and cut the roof support 3/4" tall. Crosscut the piece to 9" long. Measure in 1/2" from each edge along the base. Drill the 3/16" diameter hole through the support. Drill the counter bore 3/4" diameter with a Forstner bit 1/2" deep in the bottom of the support. Glue the support to the inside of the side walls with the top surfaces flush with the top edges of the side walls.

Take one of the roof sections and place it on top of the side walls 1" from the outside of the side wall. Mark the locations of the two holes in the roof support. Remove the roof section and file a 3/32" radius slot

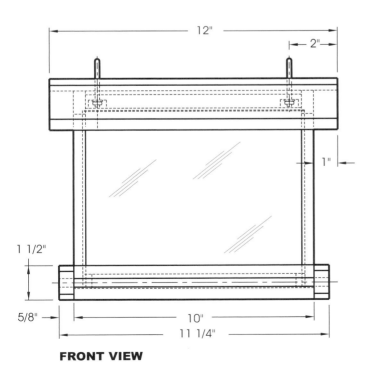

FRONT VIEW

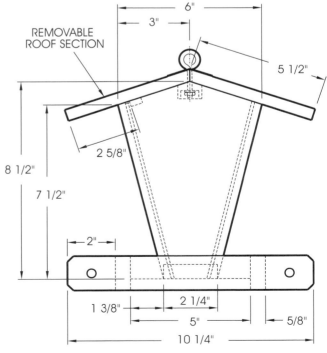

SIDE VIEW

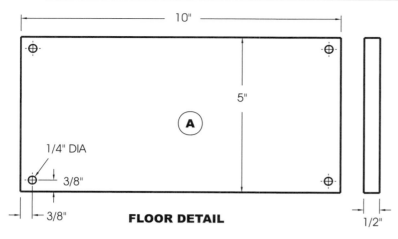

FLOOR DETAIL

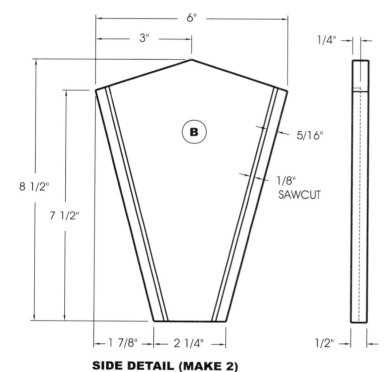

SIDE DETAIL (MAKE 2)

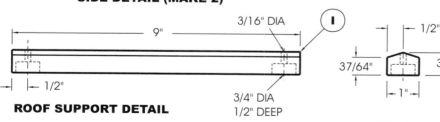

ROOF SUPPORT DETAIL

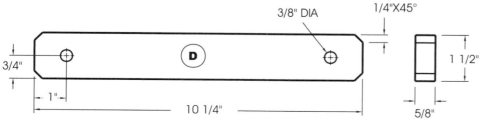

PERCH SUPPORT DETAIL (MAKE 2)

at the two marks to clear the screw eyes. Mark the other roof section and file the same 3/32" radius in this roof section also. Place the back roof on the side walls again and drill two countersunk holes through the roof and into the side walls. Screw the roof to the side walls using the 1/2" long No. 6 flat head wood screw (**P**).

Cut the roof stop (**J**) 1/2" wide by 9" long from a piece of 1/4" western red cedar. Glue it to the underside of the front roof 2 5/8" from the front edge and 1 1/2" from the side.

Cut the roof flashing (**K**) 2" wide by 12" long from a piece of galvanized metal strip. Drill the two holes to clear the screw eyes. Bend the roof flashing the same angle as the roof peak. Put the roof flashing on top of the roof and put the two eyebolts (**L**) through the holes in the roof flashing and the roof support. Add the washer (**M**) and nut (**N**). This is best done with the front glass wall removed. Now test fit the other roof section and make any adjustment necessary.

When everything fits well remove the eyebolts, roof flashing, back roof and glass for painting.

Use a non-toxic finish for the bird feeder and do not paint the inside floor, side walls nor roof of the feeder. We used an exterior latex enamel to paint our feeder. We painted our bird feeder a light tan with a light green roof. Apply two or three coats and allow to dry. When dry put the two screw eyes into the holes in the roof and add the washer and nut.

Hang the covered hopper feeder from a sturdy branch of a tree and soon you will have birds visiting your new feeder.

Open Hanging Bird Feeder

This hanging bird feeder is a lot bigger than the small simple feeder in this book and was built to allow easier access for the doves who were trying to use the smaller hanging feeder. This feeder is built of 1/2" thick construction grade plywood, redwood and western red cedar.

Cut all the wood pieces to size as listed in the list of materials. Measure in 3/8" from the edges in each corner of the floor (**A**) for the drain hole center lines. Drill the four 1/4" diameter holes.

Next lay out the two uprights (**B**) using the upright detail drawing. Using a band saw cut away from the lines and then sand to the lines. Lay out the outline of the two upright bases on the underside of the floor. Drill two No. 8 countersunk screw holes in the floor at each of the base outlines. Transfer the dimensions of one of the holes to each of the upright bases and drill a pilot hole into each upright. Attach each upright to the base using the 1 1/4" long No. 8 flat head wood screw (**K**).

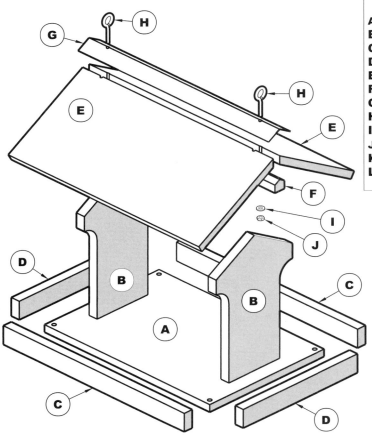

	LIST OF MATERIALS		
	DESCRIPTION	MATERIAL	FINISHED SIZE
A	FLOOR	PLYWOOD	1/2 x 9 x 13
B	UPRIGHT (2)	RED CEDAR	3/4 x 6 x 9
C	SIDE EDGING (2)	REDWOOD	5/8 x 1 1/4 x 6 1/2
D	FRONT EDGING (2)	REDWOOD	5/8 x 1 1/4 x 8 1/2
E	ROOF (2)	PLYWOOD	1/2 x 6 1/2 x 15
F	ROOF SUPPORT (2)	RED CEDAR	3/4 x 3/4 x 2
G	ROOF COVERING	CLOTH TAPE	1 1/2 WIDE x 15
H	EYEBOLT (2)	STEEL	3/16 x 2 GALV.
I	WASHER (2)	STEEL	No 10 GALV.
J	NUT (2)	STEEL	10-32 GALV.
K	WOOD SCREW (8)	BRASS	1 1/4 x No 8 FL HD
L	BOX NAIL (20)	STEEL	4d GALVANIZED

Position the two uprights over the other holes. Drill the pilot hole in the upright using the other hole in the base as a guide. Remove the wood screw and set the upright aside for now.

Drill pilot holes in the two side edgings (**C**) and the two front edgings (**D**) for the 4 penny box nails (**L**). Glue and nail the front edging and the side edging pieces to the edge of the floor.

Sand the bevel to the top edge of the two roofs (**E**). Cut the angle on

OPEN HANGING BIRD FEEDER ASSEMBLY

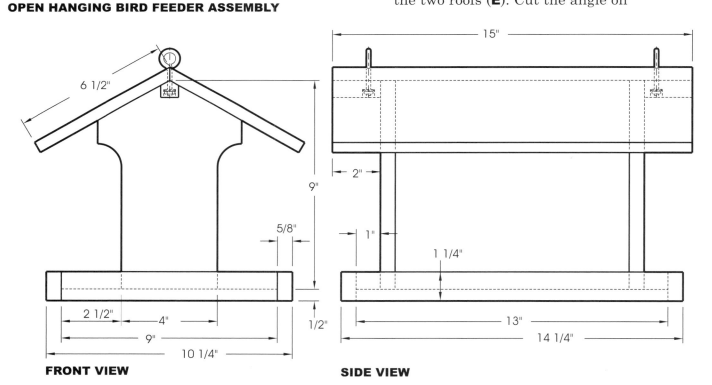

FRONT VIEW

SIDE VIEW

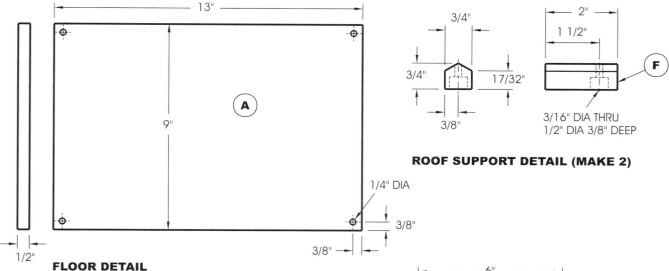

FLOOR DETAIL

ROOF SUPPORT DETAIL (MAKE 2)

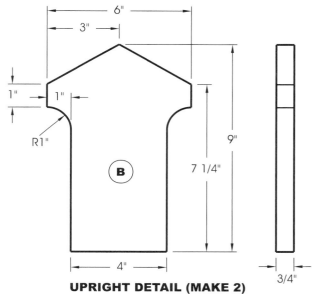

UPRIGHT DETAIL (MAKE 2)

the top of the two roof supports (**F**). Drill the 3/16" diameter hole through the support and drill the 1/2" diameter counterbore 3/8" deep.

It is now time to do the final assembly of the feeder. Glue and screw the two uprights to the floor. Apply glue to the two roof section bevels and to the two upright peaks. Place the roof sections on the uprights with a 2" overhang on both sides and put masking tape across the peak and allow to dry.

When the glue dries for the roof, remove the masking tape and lay out the center of the screw holes on the roof for the uprights. Drill the four screw holes with a counter sink combination drill. Screw the screws into the uprights and apply wood putty to cover the holes.

Measure in 1 1/2" from each edge along the peak. Drill through the peak with a 3/16" diameter drill. Glue the supports to the underside of the roof and the upright. The two 3/16" diameter holes should line up, if not make any adjustments necessary.

Put a piece of roof covering (**G**) along the peak of the roof. With the point of an awl poke a hole through the tape.

We painted our bird feeder a light tan with a light green roof. Apply two or three coats and allow to dry. Do not paint the floor of the feeder. We left the uprights and the edging unpainted so it will weather to a natural warm brown over time. When dry put the two eyebolts (**H**) into the holes in the roof and add the washers (**I**) and nuts (**J**).

Hang the open hanging feeder from a sturdy branch of a tree and soon you will have birds visiting your new feeder.

Multi Station Bird Feeder

The multi station bird feeder has twelve feeding trays with perches for the birds to get their food. You can tell when the feeder is getting low on birdseed when all the birds gather at the lower feed positions. The feeder hangs from a tree limb and the top cap is removable for easy filling of birdseed without removing the feeder from the tree. The walls are made of 1/2" thick construction grade plywood with the rough side out. The small roofs are made from 1/4" thick plywood and the other pieces are made of redwood.

Cut a piece of 1/2" thick plywood 16" wide by 31" long. Set the fence on your table saw and rip three pieces 3 1/2" wide and two pieces 2 1/2" wide. Take one of

Multi Station Bird Feeder

the 3 1/2" wide plywood pieces and crosscut two caps (**A**) 3 1/2" long. Take two of the 3 1/2" wide pieces and lay out the hole center lines for the two fronts (**B**). Take two of the 2 1/2" wide pieces and lay out the hole center lines for the two sides (**C**). Drill all of the 3/4" diameter feed holes and all of the 1/4" diameter perch holes. Cut out the inner cap (**D**) 2 1/2" by 2 1/2" from the remaining section of plywood.

Make the feed inner cap (**E**) from a piece of 1 1/2" by 3 1/2" redwood. Cut the block 1 1/8" tall by 2 1/2" wide and 2 1/2" long. Layout the peak 5/8" up from the bottom

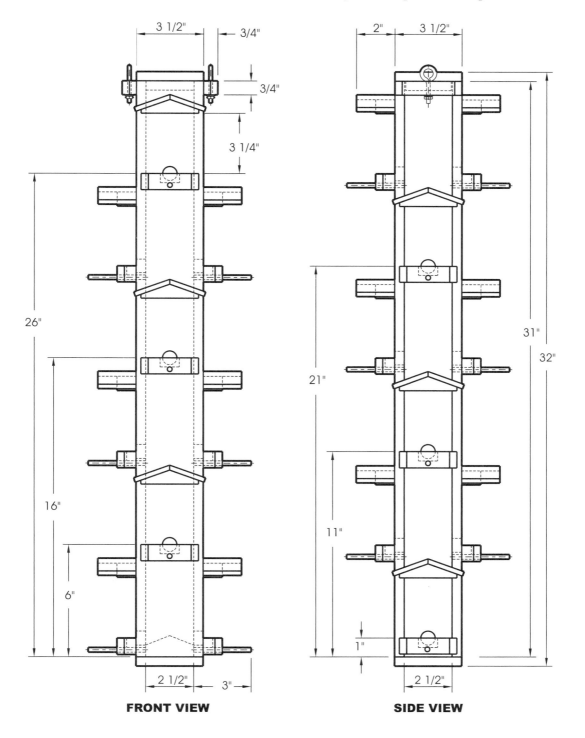

FRONT VIEW **SIDE VIEW**

on both sides and 1 1/4" in from the side. Sand the block to the lines.

Lay out the twelve roof supports (**G**) 1" wide by 3" long on a piece of 3/4" thick redwood. Cut the pieces on a band saw and sand the edges smooth. Layout the peak 7/32" up from the bottom on both sides and 1 1/2" in from the side. Sand the block to the lines.

Rip two 25" long pieces of 1/4" thick plywood 2" wide for the twenty-four roofs (**F**). Cross cut to 2" long. Set the table on a bench sander to the angle of the roof support and sand the top edge of the roof pieces.

Rip a 20" long piece of 1 1/2" by 3 1/2" redwood 7/8" high by 2 1/8" wide. Cross cut 6 pieces 3" long for the twelve feed blocks (**H**). Lay out the centers for the feed holes and the perch holes. Drill the

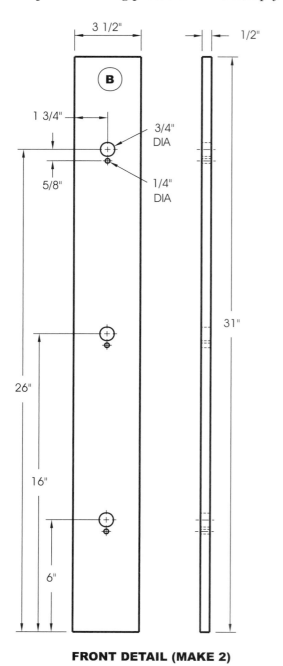

FRONT DETAIL (MAKE 2)   **SIDE DETAIL (MAKE 2)**

Multi Station Bird Feeder

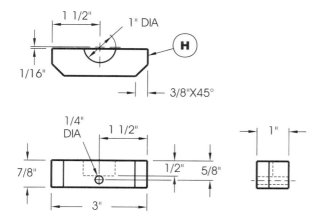

FEED BLOCK DETAIL (MAKE 12)

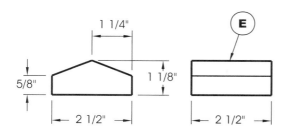

FEED INNER CAP DETAIL

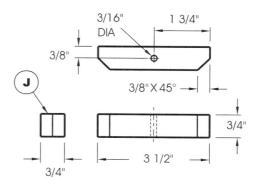

HANGING BLOCK DETAIL (MAKE 2)

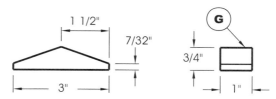

ROOF SUPPORT DETAIL (MAKE 12)

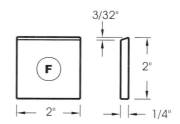

ROOF DETAIL (MAKE 24)

1/4" diameter perch hole through the blocks. Drill the 1" diameter feed hole 1/2" deep with a Forstner bit. Cut each block down the center making two identical pieces each 1" wide. Lay out the 3/8" by 45 degree chamfers on the two corners of each block. Sand the edges.

Lay out the two hanging blocks (**J**) 3/4" wide by 3 1/2" long on a piece of 3/4" thick redwood. Cut the pieces on a band saw and sand the edges smooth. Lay out the center of the hole for the eyebolt and the two 3/8" by 45 degree chamfers. Drill the 3/16" diameter hole. Cut the chamfers and sand smooth.

It is easier to add the roof supports, feed blocks and hanging blocks before the walls are assembled. Lay out the position of the roof supports and the feed blocks on the walls.

Glue and nail, using the 4 penny box nails (**N**), the roof supports to the walls from the inside. Drill two pilot holes through the walls at the support locations. Place the roof supports to the wall and drill the pilot holes three-quarters of the way into the roof support.

Slip a scrap piece of dowel into the hole in the feed block and wall. Drill two pilot holes for each feed block and then glue and nail from the inside the feed blocks to the walls.

Glue and nail the two hanging blocks to the side walls with the holes facing up.

Glue and nail the front wall to the edge of the side walls. Then glue and nail the other front wall to the edge of the side walls. Next, glue and nail the feed inner cap to the walls. Glue and nail the lower cap to the bottom edges of the walls. Glue the roofs to the top of the roof supports.

Glue the perches (**I**) into the holes in the feed block and into the hole in the walls.

Check the fit of the inner cap inside the top opening of the walls. We want a snug fit so the cap will not come off when the wind blows. Sand the edges of the inner cap as needed to get the correct fit. Then glue and nail the top cap to the inner cap.

Use a non-toxic finish for the bird feeder. We used an exterior latex enamel to paint our feeder. Paint the feeder your favorite color and the roof a contrasting color. We painted the exterior walls a light brown and the top cap a light green. The hanging blocks, feed blocks and perches were left unpainted and will weather to a warm brownish red color over time.

Insert the two eyebolts (**K**) into the holes in the top and add the washer (**L**) and nut (**M**). Put a small dab of white glue on the threads to keep the nut from coming loose.

Hang the multi station feeder in a tree in your yard and wait for the birds to come calling. When the birds start feeding only at the lower feeder perches, you will know it is time to refill the feeder. **tp**

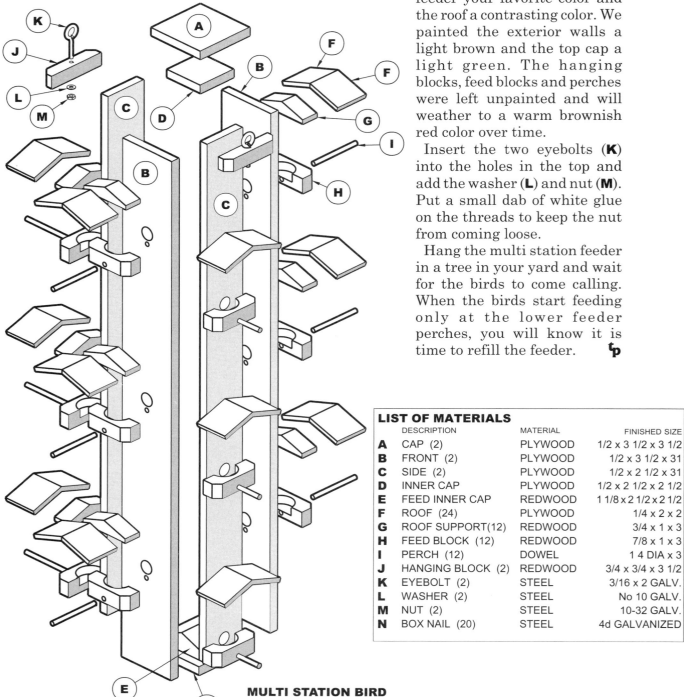

MULTI STATION BIRD FEEDER ASSEMBLY

LIST OF MATERIALS

	DESCRIPTION	MATERIAL	FINISHED SIZE
A	CAP (2)	PLYWOOD	1/2 x 3 1/2 x 3 1/2
B	FRONT (2)	PLYWOOD	1/2 x 3 1/2 x 31
C	SIDE (2)	PLYWOOD	1/2 x 2 1/2 x 31
D	INNER CAP	PLYWOOD	1/2 x 2 1/2 x 2 1/2
E	FEED INNER CAP	REDWOOD	1 1/8 x 2 1/2 x 2 1/2
F	ROOF (24)	PLYWOOD	1/4 x 2 x 2
G	ROOF SUPPORT(12)	REDWOOD	3/4 x 1 x 3
H	FEED BLOCK (12)	REDWOOD	7/8 x 1 x 3
I	PERCH (12)	DOWEL	1 4 DIA x 3
J	HANGING BLOCK (2)	REDWOOD	3/4 x 3/4 x 3 1/2
K	EYEBOLT (2)	STEEL	3/16 x 2 GALV.
L	WASHER (2)	STEEL	No 10 GALV.
M	NUT (2)	STEEL	10-32 GALV.
N	BOX NAIL (20)	STEEL	4d GALVANIZED

Center Column Bird Feeder

This center column bird feeder will hold quite a bit of bird seed and with the generous sized platform will feed a good number of birds at one time. The removable roof allows filling from the top. Four openings at the base of the walls allow a small but adequate amount of seed to spill out on to the platform. This is a good way to conserve the seed. The birds do not see a lot of seed and therefore do not spill as much to the ground. This feeder is built from 1/2" thick construction grade plywood, redwood and western red cedar for the roof support.

Cut the floor (**A**) 12" wide by 12" long from a piece of 1/2" thick plywood. Lay out the center of the four drain holes in each corner 3/8" from the edges. Drill the 1/4" diameter holes.

Rip two 12 1/2" long pieces of 1/2" thick plywood 4 1/2" wide for the two fronts (**B**). Lay out the peak on the two front walls and the 1 1/2" wide by 1/2" tall openings at the bottom using the dimensions from the front detail drawing. Cut the openings and the roof peaks. Lay out the center of the dowel hole. Drill the 1/4" diameter hole through the front.

Then rip two 11 1/2" long pieces of 1/2" thick plywood 3 1/2" wide for the sides (**C**). Lay out and cut the openings at the bottom of the two side walls. Tilt the table saw blade to match the slope of the roof peak of the front wall and crosscut the top of the side walls making the lower (outside edge) 11" long.

Glue and nail, using the 4 penny box nails (**J**), the front walls to the edges of the side walls.

Rip four 13" long pieces of 1 1/2" thick redwood 1/4" wide for the edging. Crosscut two pieces 12" long for the side edges (E) and two pieces 12 1/2" long for the front edges (**D**). Glue and nail the edging to the edge of the floor. Drill pilot holes in the redwood to prevent splitting.

Cut two pieces of 1/2" thick plywood 8" wide by 13" long for the two roofs (**F**). Tilt the table saw blade to match the slope of the roof peak of the front wall and crosscut the top of the two roof pieces making the lower (outside edge) 7 1/2" long. Put a bead of glue on each beveled edge of the roof pieces and put one on the walls lining up the bottom edge of the roof bevel with the peak of the front walls. Place the other roof section against the first and add two pieces of masking tape across the peak to hold the roof sections together while the glue dries.

With the table saw blade still set to the roof angle cut the top of the roof block (**G**) from a piece of 1 1/2" thick western red cedar. Make the overall height 1" and cut the piece 3 1/4" wide by 3 1/4" long. Lay out the center of the dowel hole. Drill the 1/4" diameter hole through the support. Remove the roof from the walls and glue the roof support block to the bottom of the roof in the center.

Place the wall assembly in the center of the base and trace around the walls on to the floor. Also mark the location of the openings. Then measure in 1/4" from the first line and drill two pilot holes for the nails. Keep away from the notches in the wall. Do this for all four walls. Glue and nail the wall assembly to the floor.

Add the roof cover (**H**) to the peak of the roof. Burnish the tape to improve the contact to the rough plywood surface. Then lightly sand the surface of the tape so the

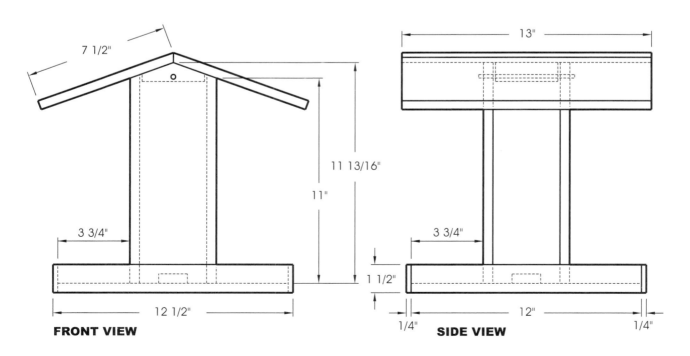

FRONT VIEW   **SIDE VIEW**

Center Column Bird Feeder

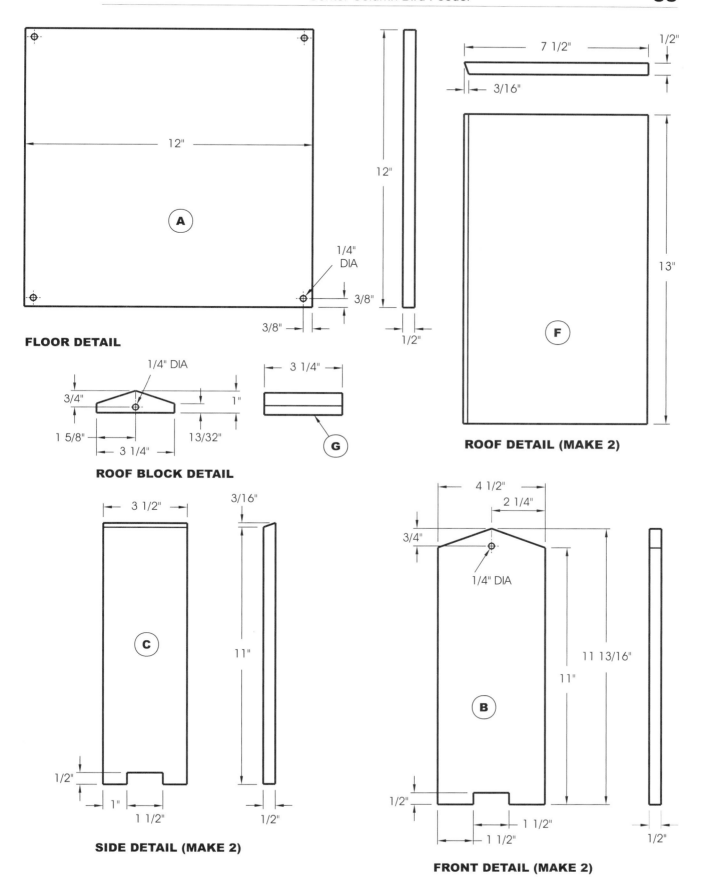

54 Simple Birdhouses and Feeders of Wood

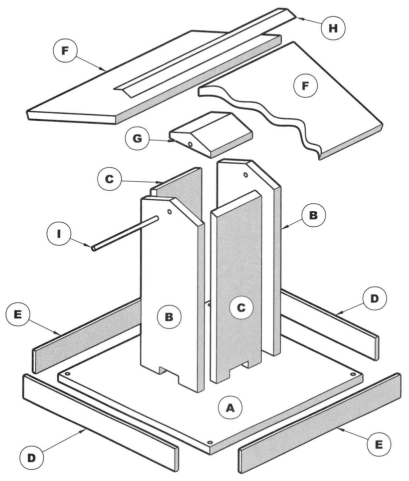

CENTER COLUMN BIRD FEEDER ASSEMBLY

LIST OF MATERIALS

	DESCRIPTION	MATERIAL	FINISHED SIZE
A	FLOOR	PLYWOOD	1/2 x 12 x 12
B	FRONT (2)	PLYWOOD	1/2 x 4 1/2 x 11 13/16
C	SIDE (2)	PLYWOOD	1/2 x 3 1/2 x 11 3/16
D	FRONT EDGE (2)	REDWOOD	1/4 x 1 1/2 x 12 1/2
E	SIDE EDGE (2)	REDWOOD	1/4 x 1 1/2 x 12
F	ROOF (2)	PLYWOOD	1/2 x 7 1/2 x 13
G	ROOF BLOCK	RED CEDAR	1 x 3 1/4 x 3 1/4
H	ROOF COVER	CLOTH TAPE	1 1/2 WIDE x 13
I	PEG	DOWEL	1/4 DIA x 5
J	BOX NAILS (20)	STEEL	4d GALVANIZED

paint will adhere better.

Use a non-toxic finish for the bird feeder and do not paint the inside surface of the edging or the floor of the feeder. We used an exterior latex enamel to paint our feeder. Paint the feeder your favorite color and the roof a contrasting color. Paint the underside of the roof, edging and the walls a light color and the roof and the roof edges a darker color. We painted our bird feeder a light tan with a light green roof. Apply two or three coats and allow to dry. When dry put the roof on the walls and slide the peg (**I**) through the holes in the walls and roof support.

Mount the center column feeder on a block on top of a sturdy post and soon you will have birds visiting your feeder. **tp**

Bird Water Platform

In addition to feeding our little feathered friends, they need water too. They need water to drink and to bathe in. The water platform can be as simple or as elaborate as you want to make one. A concrete birdbath can be built if you have the talent and the room to put one in your yard. Our water platform is a bit more on the modest side. We used a 14" diameter terracotta dish as the starting point for our bird water platform. The platform is 1/2" thick construction grade plywood and the other pieces are redwood and western red cedar.

Cut the parts as specified in the list of materials. Lay out the centers for the large hole and the eight countersunk wood screw holes in the floor (**A**). With a compass lay out the 12" diameter hole in the center. Cut the hole with a portable jigsaw and sand the edges smooth with a drum sander. Drill the countersunk holes.

Drill pilot holes in the two front edgings (**B**) and the two side edgings (**C**). Glue and nail the edging pieces to the edge of the floor using the 4-penny box nails (**I**).

Lay out the dish relief and angle on the two supports (**D**). Drill the 1" diameter hole for the 1/2" radius in the corner of the relief and cut the lines with a band saw. Sand the edges smooth.

Lay out the dish relief and angle on the four half supports (**E**). Drill the 1" diame-

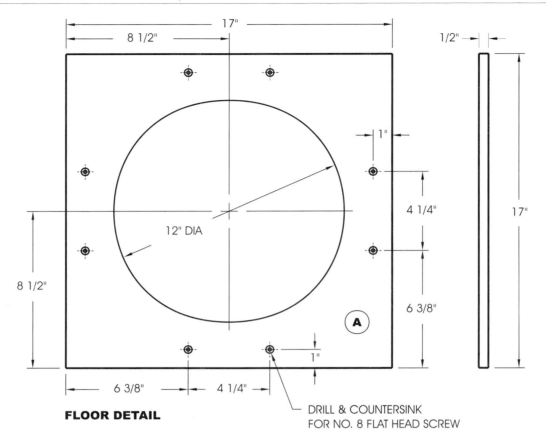

FLOOR DETAIL

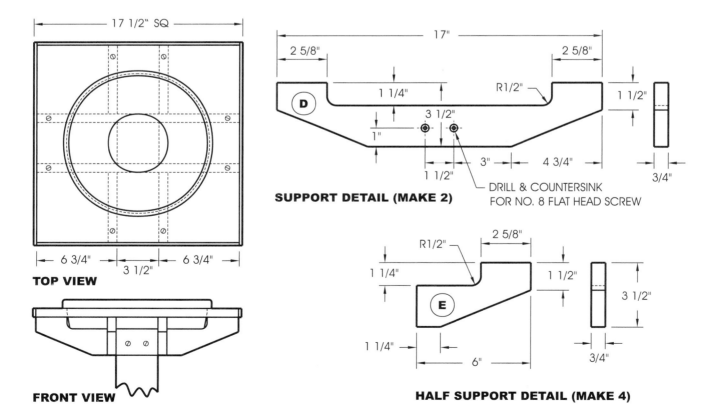

TOP VIEW

FRONT VIEW

SUPPORT DETAIL (MAKE 2)

HALF SUPPORT DETAIL (MAKE 4)

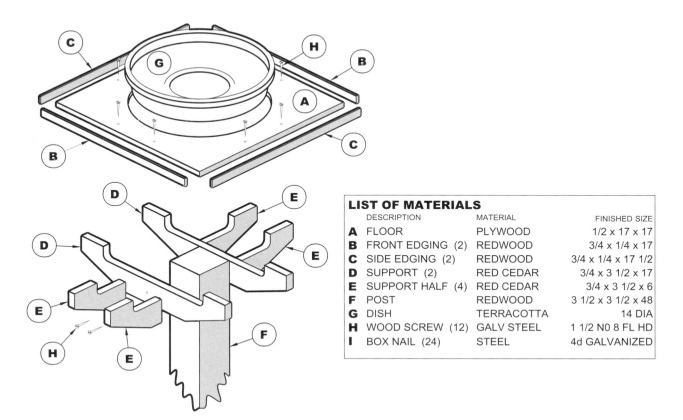

BIRD WATER PLATFORM ASSEMBLY

LIST OF MATERIALS

	DESCRIPTION	MATERIAL	FINISHED SIZE
A	FLOOR	PLYWOOD	1/2 x 17 x 17
B	FRONT EDGING (2)	REDWOOD	3/4 x 1/4 x 17
C	SIDE EDGING (2)	REDWOOD	3/4 x 1/4 x 17 1/2
D	SUPPORT (2)	RED CEDAR	3/4 x 3 1/2 x 17
E	SUPPORT HALF (4)	RED CEDAR	3/4 x 3 1/2 x 6
F	POST	REDWOOD	3 1/2 x 3 1/2 x 48
G	DISH	TERRACOTTA	14 DIA
H	WOOD SCREW (12)	GALV STEEL	1 1/2 N0 8 FL HD
I	BOX NAIL (24)	STEEL	4d GALVANIZED

ter hole for the 1/2" radius in the corner of the relief and cut the other cuts with a band saw. Sand the edges smooth.

Clamp the two supports to the post (**F**) with the top edges of the support in line with each other. Place the floor on top of the supports with 6 3/4" from the inside surface of the support to the outside edge of the floor from both ends. Drill the pilot holes into the supports using the countersunk holes in the floor as a guide. Add four flat head screws (**H**). Remove the clamp and place the floor and support assembly upside down on your workbench. Mark a line 6 3/4" from each end of the support. Glue the four half supports to the support and allow to dry. Drill the pilot holes into the half supports using the countersunk holes in the floor as a guide. Add the four wood screws. Remove all the wood screws and nail the supports to the half supports. Screw the support assembly to the floor.

Place the floor/support assembly on the post and drill pilot holes into the post using the countersunk holes in the supports as a guide. Add the four wood screws.

Paint the entire assembly with several coats of Thompson's water sealer, or any similar product. With water splashing on the deck it will have to be sealed often throughout the years.

Find a cool place in your backyard and plant the post in the ground. Use concrete if desired for a permanent installation. Place the dish (**G**) into the hole in the platform and add water and your feathered friends will thank you.

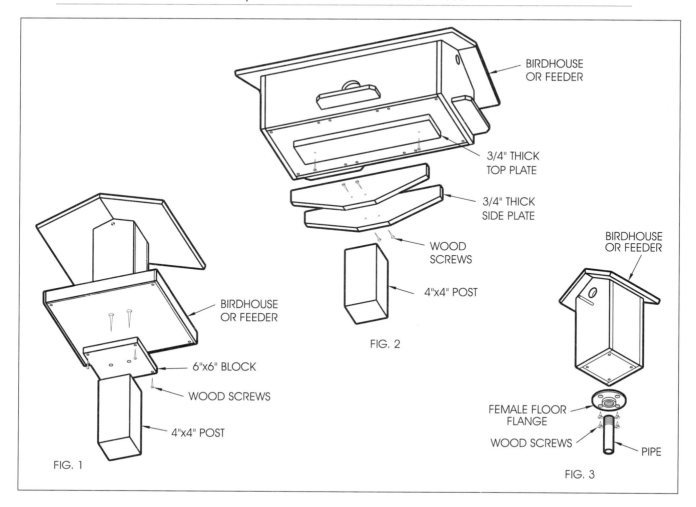

FIG. 1
FIG. 2
FIG. 3

Mounting and Poles

There are many ways to install birdhouses and feeders outdoors. They may be hung from a tree limb; nailed to the side of a building, shed or barn; or mounted on top of a pole; or nailed to the side of a post. The design of the birdhouse or feeder will often dictate how the structure is mounted.

To hang a birdhouse or feeder from a tree limb some type of fastener must be added to the roof. We used eyebolts for several of our birdhouses and feeders. This type of fastener is very good since it traps the wood between the eye head and the nut. On the other hand, screw eyes are not acceptable since over time they may pull away from the roof causing the birdhouse to fall to the ground. Other methods are also good, for example a bent nail or hole drilled through the roof under the eves will also work. Rope, chain or wire can be used to hang the structures. The squirrels in our neighborhood are too smart for the chains or rope, they simply pull the bird feeder up to the limb and feast on the birdseed. We have since settled on wire as the only squirrel proof material for our yard.

A couple of our birdhouses and feeders were designed to be nailed to the side of a

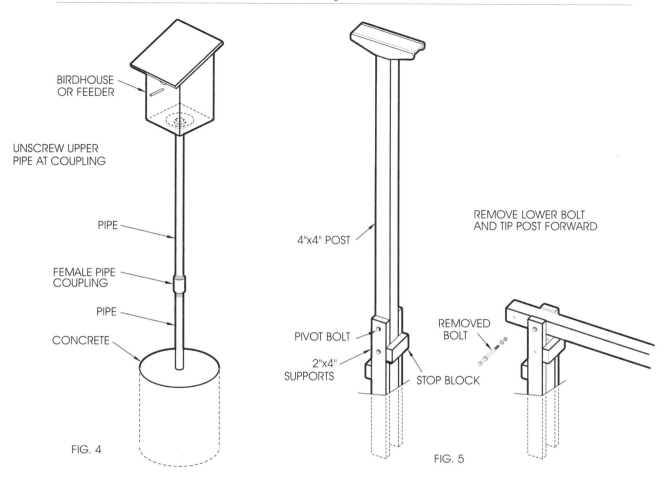

FIG. 4

FIG. 5

barn, shed or post. A nailer support was added to the back of these structures with holes for clearance of the nails or screws that will secure the birdhouse or feeder to the siding or post.

If the birdhouse or feeder can not be hung from a tree or nailed to the side of a building then the best way to mount it is on a pole. The size of the pole and the method of mounting will depend on the size and weight of the birdhouse or feeder. A wood post in the ground can be used to mount a feeder or house. For a small light structure nail or screw a 3/4" thick piece of plywood, about 1 inch larger than the post all around, to the top of a post. Drill four holes in the block and screw the structure to the block (Fig 1). A larger structure will need more support than just the simple block. The top block can be made longer and two supports added to the side of the block and post (Fig 2). For a small feeder or house a female floor flange can be attached to the bottom of the structure (Fig 3). The pipe flange can be screwed to the top of a pipe sticking out of the ground. Iron or pvc pipe could be used.

The birdhouse must be accessible for cleaning at least once a year. A stepladder could be used and the birdhouse either removed or cleaned on the spot. The pole could be made to come apart and the upper part, that is attached to the birdhouse, lowered for cleaning. If threaded pipe is used for the mounting pole a coupling can be added near the ground and the top pipe is simply unscrewed from the coupling and the birdhouse lowered to the ground (Fig 4). The pole could also be hinged to allow the upper section with the birdhouse lowered to the ground. A wood post is ideal for this type of design with a single bolt used as the hinge and another bolt added to secure the pole in the upright position (Fig 5).

Predator Guards

Once the birdhouse or feeder is built and mounted in the backyard, you will soon see the need to devise a way to keep out that pesky squirrel or other animals. Several different setups can be used to keep the predators from the birdhouses or feeders.

Hanging birdhouses and feeders:

If the birdhouse or feeder is hung from a tree branch or hanging post, old discarded pie tins can be used to keep the pesky squirrels away. Punch a hole in the center of three pie tins. Cut two sections of old garden hose about 3" long. String the pie tins and the hose sections as shown in Fig 1. Tie one end around the tree limb and the other to the structure. Put a piece of hose or rubber sheet between the wire and the tree limb to protect the limb. If pie tins are not available, a thin aluminum sheet could be cut into 10" diameter disks.

Sheet metal cone:

Sheet metal can be cut and formed into a cone (Fig 2) and for suspended birdhouses can be placed above the structure to keep the squirrel out of the birdhouse or feeder. Make the hole in the center large enough so the cone will tip from side to side. An inverted cone can also be used for a pole mounted birdhouse or feeder (Fig 3). This cone should be larger that the one mounted above the structure and the hole made so there is plenty of clearance.

Sheet metal stovepipe:

A piece of aluminum sheet can be rolled into a cylinder and put around a pole or post (Fig 4). Join the ends with sheet metal screws or blind rivets. The bottom of the cylinder should be at least five feet above ground level. Attach the cylinder to the post with two to four 10 or 16-penny nails. Stovepipe of the proper diameter could also be substituted for the metal cylinder.

Sheet metal for a square post:

For a large square post for the larger birdhouses a four-sided pyramid can be built of sheet metal (Fig 5) and nailed to the post. The thinner the sheet the better. You do not want the guard so sturdy that it will support the cat or squirrel. Make sure the bottom of the sheet metal is at least four feet from the ground.

This is a small sampling of simple neat predator guards that can be used on a host of birdhouses and feeders. However, be prepared for that pesky squirrel who is able to out smart these simple designs. He may need something special and more elaborate.

tp

Preditor Guards

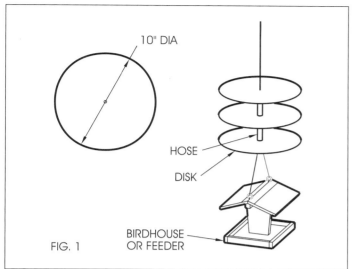

FIG. 1

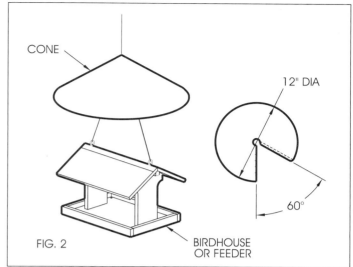

FIG. 2

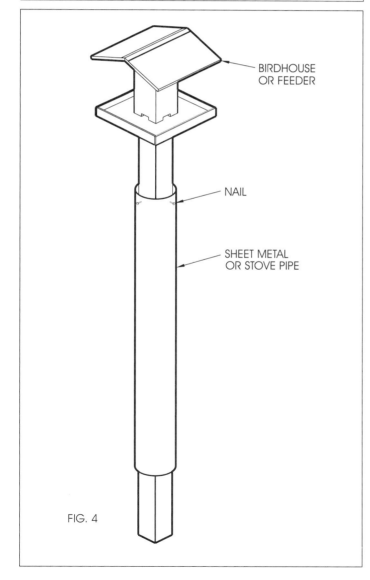

FIG. 4

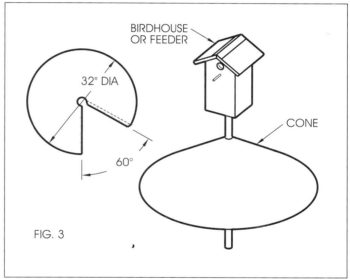

FIG. 3

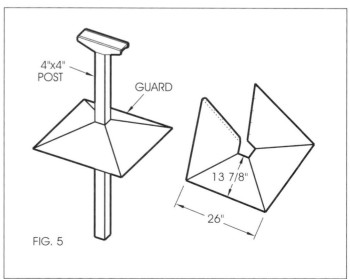

FIG. 5

Birdhouse Size Specifications

Types	Entrance Diameter (inches)	Entrance above Floor (inches)	Interior Height of House (inches)	Floor of House (inches)	Height above Ground (feet)
Bluebird					
Eastern	1.5	6	8	5 x 5	5 to 10
Mountian	1.5	6	8	5 x 5	5 to 10
Western	1.5	6	8	5 x 5	5 to 10
Chickadee					
Boreal	1.125	6 to 8	8 to 10	4 x 4	6 to 15
Black-capped	1.125	6 to 8	8 to 10	4 x 4	6 to 15
Carolina	1.125	6 to 8	8 to 10	4 x 4	6 to 15
Chestnut-backed	1.125	6 to 8	8 to 10	4 x 4	6 to 15
Gray-headed	1.125	6 to 8	8 to 10	4 x 4	6 to 15
Finch					
House	2	3 to 4	6	6 x 6	8 to 12
Purple	2	3 to 4	6	6 x 6	8 to 12
Flicker	2.5	14 to 16	16 to 18	7 x 7	6 to 20
Flycatchers					
Great-crested	2	6 to 8	8 to 10	6 x 6	8 to 20
Olivaceous	2	6 to 8	8 to 10	6 x 6	8 to 20
Western	2	6 to 8	8 to 10	6 x 6	8 to 20
Kestrel	3	9 to 12	12 to 15	8 x 8	10 to 30
Martin	2.5				
Purple	2	1	6	6 x 6	15 to 20
Western	2	1	6	6 x 6	15 to 20
Nuthatch					
Brown-headed	1	6 to 8	6 to 8	2 X 3	5 to 20
Red-breasted	1.25	6 to 8	6 to 8	4 X 4	5 to 20
White-breasted	1.25	6 to 8	6 to 8	4 X 4	5 to 20
Owl					
Barn	6	4	15 to 18	10 x 18	12 to 18
Barred	8		16	13 x 15	10 to 30
Screech	3	9 to 12	12 to 15	8 x 8	10 to 30
Phoebes					
Black	one side open		6	6 x 6	8 to 12
Eastern	one side open		6	6 x 6	8 to 12

Birdhouse Size Specifications

Types	Entrance Diameter (inches)	Entrance above Floor (inches)	Inerior Height of House (inches)	Floor of House (inches)	Height above Ground (feet)
Pigeon	4	4	8	8 x 8	10
Sparrows					
House	1.5	6	15	10 x 10	10 to 20
Song	all sides open		8 to 10	6 x 6	1 to 3
Swallows					
Barn	open one side		6	6 x 6	8 to 10
Tree	1.5	1 to 5	6	5 x 5	10 to 15
Violet-green					
Thrushes (Robin)	open one side		8	6 x 8	6 to 15
Titmouse					
Bridled	1.25	6 to 8	8 to 10	4 x 4	6 to 15
Plain	1.25	6 to 8	8 to 10	4 x 4	6 to 15
Tufted	1.25	6 to 8	8 to 10	4 x 4	6 to 15
Warbler	1.5	5	8	4 x 4	4 to 7
Wood duck	4	18 to 20	24 to 26	11 x 11	10 to 25
Woodpecker					
Downy	1.25	6 to 8	8 to 10	4 x 4	6 to 20
Flicker	2.5	14 to 16	16 to 18	7 x 7	6 to 20
Golden front	2	9 to 12	12 to 15	6 x 6	12 to 20
Hairy	1.5	9 to 12	12 to 15	6 x 6	12 to 20
Pileated	3	10 to 12	12 to 30	8 x 8	12 to 60
Red-bellied	2.5	10 to 12	12 to 15	6 x 6	12 to 20
Redheaded	2	9 to 12	12 to 15	6 x 6	12 to 20
Wren					
Bewick's	1	1 to 6	6 to 8	4 x 4	6 to 10
Broen-throated	1	1 to 6	6 to 8	4 x 4	6 to 10
Carolina	1.125	1 to 6	6 to 8	4 x 4	6 to 10
House	1	1 to 6	6 to 8	4 x 4	6 to 10
Winter	1 x 2 1/2	4 to 6	6 to 8	4 x 4	5 to 10

Index

B
Birdhouse
  Bluebird Birdhouse, 17
  Carolina Wren Tower Birdhouse, 12
  Chickadee Birdhouse, 27
  Typical Birdhouse, 5
  Western Martin Triplex Birdhouse, 21
  Wren Birdhouse, 9
Birdhouse Size Specifications, 62
Bird Feeder
  Center Column Bird Feeder, 51
  Covered Hopper Bird Feeder, 39
  Hopper Wall Mounted Bird Feeder, 34
  Multi Station Bird Feeder, 46
  Open Hanging Bird Feeder, 43
  Small Hanging Bird Feeder, 31
Bird Water Platform, 55
Bluebird Birdhouse, 17

C
Carolina Wren Tower Birdhouse, 12
Center Column Bird Feeder, 51
Chickadee Birdhouse, 27
Covered Hopper Bird Feeder, 39

F
Feeder
  Center Column Bird Feeder, 51
  Covered Hopper Bird Feeder, 39
  Hopper Wall Mounted Bird Feeder, 34
  Multi Station Bird Feeder, 46
  Open Hanging Bird Feeder, 43
  Small Hanging Bird Feeder, 31

G
Guards, Predator, 60

H
Hopper Wall Mounted Bird Feeder, 34

M
Mounting and Poles, 58
Multi Station Bird Feeder, 46

O
Open Hanging Bird Feeder, 43

P
Predator Guards, 60

S
Small Hanging Bird Feeder, 31

T
Typical Birdhouse, 5

W
Western Martin Triplex Birdhouse, 21
Wren Birdhouse, 9
Wren, Carolina, Tower Birdhouse, 12